Published by Bob Tuck
Low Worsall, Yarm,
North Yorkshire, England TS15 9QA

ISBN 0 9521938 5 X

First Published 2001

Copyright 2001 Bob Tuck

Other books by Bob Tuck

Moving Mountains
Mountain Movers
Mammoth Trucks
Hauling Heavyweights
The Supertrucks of Scammell
*Move It (compendium of Moving Mountains
and Mountain Movers)*
Carrying Cargo
Classic Hauliers
Robsons
Classic Hauliers 2
The Golden Days of Heavy Haulage
A Road Transport Heritage
A Road Transport Heritage Vol II
A Road Transport Heritage Vol III
100 Years of Heavy Haulage
Trucks (Reprint of Mammoth Trucks)
King of the Road

Printed in Great Britain by
The Amadeus Press Ltd
Cleckheaton,
West Yorkshire

Typesetting by Highlight Type Bureau Ltd
Bradford, West Yorkshire

Book design by
Sylbert Productions
Pavey Ark.

ACKNOWLEDGEMENTS

Jack Hill has been researching the background to this book for most of his adult life and obviously a large number of people have given him a great deal of time and consideration in tracing details of the early Hill family. His apologies to you if you are not specifically mentioned in this part of the book but rest assured, your help has made this book possible.

While Jack has taken the bulk of the photographs in use, contributions from other quarters - especially Roy Wiseman - are also an integral part of the book. The local newspapers of The Southampton Echo and Portsmouth Evening News have often featured the photographic activities of Hills of Botley to enthral their readers, so their permission to re-use some of their archive material is also appreciated.

The photographic expertise of Mike Insall, Dave Lee and Dave Weston are on show while the aerial pictures courtesy of Southern Newspapers give a fantastic over view to the changing developments of the Hill sites.

For my part, I would like to thank Phil Moth, Paul Hancox, John Harrington, George Baker, Jim Wilkinson and Malcolm Wilford, who continue to help with the most exacting of details whenever requested. Angela Swaine at the Highlight Type Bureau gave me time & consideration beyond the call of duty, while Jim Purkis was also kind enough to recall his long days at Hills (he started there in 1943).

Both Jack and Chris Hill have endeavoured to answer my strange enquiries but I must mention Jack's wife Mollie, who makes an excellent cup of milky coffee - creating inspiration when it's needed.

The quality of the finished book lies in the hands of the team from Amadeus Press. They now have the joy of working in brand new, custom-built surroundings but their efforts are also highly thought of.

Last word again goes to my wife Sylvia. How she puts up with my moods and bouts of silences ('I'm in deep thought darling') whenever I'm busy with a book beats me. I doubt if I would have the patience to reciprocate and I doubt if I could achieve what I do without her help.

HILLS OF BOTLEY

JACK HILL/BOB TUCK

CONTENTS

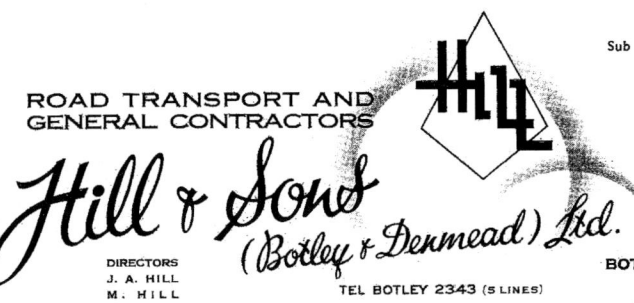

HILLS OF BOTLEY

ROAD TRANSPORT THROUGH THE CENTURIES CARRIED OUT BY THE HILL FAMILY OF HAMPSHIRE

AUTHOR'S INTRODUCTION

Travel to the southern Hampshire town of Botley today and you'll see a community clinging onto its identity. As the massive conurbation surrounding Southampton, Fareham and Portsmouth grows ever larger, Botley appears to be the last stop guardian of the countryside beyond it.

It was this peculiar mix of urban and rural influences, which has moulded Hills of Botley through more than 100 years of transport history. Originally it was the farming community and the daily demands of moving milk that introduced the Hill family to the joys of haulage. While for the last 50 years or so, Hills of Botley has been a name synonymous with abnormal loads and the demands of heavy haulage.

At their peak in the early 1960s, the company operated some 160 vehicles and trailers on a huge variety of traffic from a number of depots. Today their operation is just a fraction of that size but the fact that Hills of Botley still exist – while many of their one time competitors are no longer trading – gives some indication of their determination to continue serving the haulage industry.

Jack Hill has been part of the Hill history since he was born in 1912. It's fortunate that in his time, he has squirreled away all sorts of information while recording many of the company vehicles and their diverse activities on film. What follows is just a small slice of the company's history, which Jack would like dedicating to the memories of his grandfather George Augustus Hill, his father Valentine Augustus Hill and his elder brother Eric William George Hill.

Thanks must also be made to Jack's son Chris, who is currently ensuring that the history of Hills of Botley is a story still in the making.

Bob Tuck,
Pavey Ark, Low Worsall
Yarm, North Yorkshire
September 2001.

Seen at the top of this page in the early 1960s are JA's wife Mollie and their two sons Christopher James (CJ) - astride the Welsh mare pony 'Pinkie' - and Barry John (BJ). The habit of using the initials to denote which Hill you were talking about struck one regular caller as amusing: 'It's like speaking to the alphabetical Hills,' he said. The family are seen (above) in 1988 when Jack and Mollie were celebrating their 50th wedding anniversary. Since then, the couple have clocked up another 13 years of married life.

THE HILL FAMILY

George Augustus Hill
1852 - 1935

Valentine Augustus Hill
1882 - 1964

Eric William George Hill
1909 - 1986

John Augustus Hill
b. 1912

THE ORIGINAL FORM OF HORSE POWER

Tracing your family tree can be a fascinating exercise although you never know what you may find. Jack and Mollie Hill spent hours going through Hampshire's local records and traced their ancestors back to 1664.

Traditionally the family were farmers so it was perhaps easy enough to discount the mention of one Hill who met with an untimely demise in 1777. While working in the dockyard he set fire to the rope house and although he absconded, he was found in an alehouse in Hook near Alton. Taken back for trial, he was subsequently hung in chains from the yardarm of a ship in Portsmouth harbour.

By the 1800s, the Hill family were firmly established as pillars of the farming community. George Augustus (GA) Hill was born in 1835 and his base of operations was eventually to be at Drayton Manor Farm at Drayton (now a Portsmouth suburb).

Hill's involvement with haulage during the mid 1800s was limited to taking produce to Portsmouth market - by horse drawn market van - and using similar propulsion to carry fat farm stock to the nearest cattle market at Cosham. Drovers usually drove the stock, which was able to walk, to market but things were all set to change in 1894 when GA introduced the family to mechanical transport with the purchase of their first steam traction engine.

The power of steam opened all sorts of doors to the Hills for as well as doing haulage, all manner of ancillary equipment could be driven from the one steam engine at a much quicker pace than what horses could manage. GA's eldest son Valentine Augustus (VA) embraced this new technology when he set up at Rushmere Farm, Hambledon although not everyone welcomed steam power.

Horses from the Hill farms were still being requisitioned for use during the 1914-18 World War I as the military reckoned steamers couldn't be used near the front line as their obvious plume of smoke and sparks could too easily gave away their position. Steamers could also be a nuisance in civilian life as setting fire to your load was a regular occurrence especially if the day was hot, your load was of hay, straw or even dry timber - and you couldn't be bothered to sheet it. That's one problem hauliers of the 21st century don't have to worry about.

The horse could be the farmer's best friend and those with the right temperament could be asked to do a variety of roles. The haymaking scene at Lovedean shows the elevator being driven by horsepower and the sweeps making easy work of collection when two horses pull together.

Horses - just like humans - enjoyed to eat. With the temptation of a fresh crop so close to hand, the horse nearest the corn was fitted with a nosebag as eating the crop would have given the animal the gripes.

When not working on the farm, timber extraction was a regular job for the Hill horses either dragging the dressed down tree trunks along the ground or after they were lifted onto a trailer. When working a string like this, a light horse was always placed at the front. It was a lot easier to handle than a heavy cart horse (the others would follow their leader) although the heavier ones did most of the work. The modest looking tipcart (centre) proved a versatile load carrier. It was a regular job to employ stone pickers to clear the fields sufficiently so they could be planted with crops, although the younger Hill sons were also expected to help in this back-breaking job. Once piled up, the stones were shovelled into a tipcart and would be delivered in bulk to the nearest road-making site and used in the foundations.

Searching the records of Wallis & Steevens, the Basingstoke based engineers show that GA Hill bought his first Wallis threshing machine in August 1894 and in the following month bought a 6hp Wallis simple cylinder traction engine - makers number 2237. In September 1899 he bought a similar 7hp machine - number 2471 - and in April 1904 bought two Wallis 6 ton traction trucks. VA's purchases included a B2 Little Giant - made by W Tasker & Sons of Andover - in June 1912, a 6hp Wallis expansion gear traction engine (no. 2677) in March 1920 and a similar engine - number 7691 - in August 1921. The upper photograph shows threshing at Rookwood Farm, Denmead during World War 1. The threshing scene in the lower photograph (opposite page) is taken at Willie Parret's Broadway Farm between Denmead and Lovedean. VA Hill is seen second from the left in shirtsleeves, Ben Peters being the engine driver.

The picture of un-threshed wheat ricks and hay ricks seen on the Drayton Estate is a sight seldom seen in England today.

A big spin off for GA was the manufacture of bricks from a chalk dell on Lower Drayton Lane. Thousands of bricks were produced although being of chalk, these were used for internal walls only - like the modern day breezeblock. Horse and cart teams did the local deliveries although steam traction engines were used for long distance hauls.

Not every horseman made the transition to steam so those with the ability to handle the steam-powered engines were in great demand. Moving long distances became a way of life for some drivers and Ben Peters (VA's first driver) was employed by Hills with a letter telling him when to put his furniture on the train and a horse and cart would pick him up at Havant station.

Jack Kirby, pictured, was VA's second driver although none of the steam engine drivers made the transition to the petrol-powered vehicles. All VA's steam traction engines were sold in the 1932 Farm Auction.

VA's Fodens were purchased from John Kiln of Cosham, the main Foden agent for Hampshire. The deal was obviously helped by this interesting advertising pitch.

MAKING A FRESH START WITH MILK

Recession is a word that sends shudders through the mind of anyone involved in business but a downturn, which affected most of the country, was the one that struck in the late 1920s. The period evokes memories of the General Strike and Jarrow marchers walking to London but while the Hill problems may not have hit the same headlines, VA and his family suffered financial hard times.

As the farm and all its machinery were put up for auction in 1932, VA was loaned £500 by his father (who by then had retired) to take over the Bottings Hotel. This was initially rented from Lady Madalene Sabine Jenkins although it was later bought outright. In hindsight this proved to be a vital move for the family's future haulage development, as the 14 acres of market garden around the hotel was to later form the Bottings Industrial Estate - HQ of the Hill transport empire. Although always referred to as being at Botley - and actually situated opposite Botley railway station - this temperance hotel had the postal address of the adjacent village of Curdridge.

One thing, which made the move to Botley and wasn't sold in the auction, was Hills interest in milk haulage. The first movement of their own milk in churns was done fairly modestly with a Model T Ford and then a small Morris Commercial van. But as other farmers in the area requested Hills to move their milk, a Dennis 30cwt capacity four-wheeler (PX 7107) was bought from Sparshatts and VA's eldest son Eric William George (EWG) first drove this.

The creation of the Milk Marketing Board gave added stability to the purchase and price of all the milk then being produced. The formalised arrangements for farmers also allowed expansion of the Hill haulage interests and two other Dennis' were bought for EWG's younger brothers - John 'Jack' Augustus (JA) and Dudley Valentine (DV).

As well as moving milk in the mornings, the three Hill vehicles also did hay & straw haulage on the afternoons. A regular back load was the barley residue from both United & Buckworth Breweries which local farmers bought for cattle feed. However, pressures on the farming community were to introduce Hills to another side of business. Rather than pay for haulage & feed, the farmers agreed to barter their own produce, which Hills then sold on. This sales side to Hills was to evolve into its own department and was an important part of the company for the next 30 years.

The introduction of haulage licensing in 1933 ('A', 'B' and 'C') was to put a big value on any vehicles then in transport. As their respective owners came up to retirement, the four Hill partners expanded their fleet through buying up other people's vehicles. Although this method of growth strained the finances, it was a lot easier to acquire a vehicle - and its important carriers licence - in this fashion than by trying to get an extra vehicle through the Licensing courts.

Fred Watton's cattle wagon and milk lorry were bought up after Fred's death and similarly when Bishop's Waltham cattle haulier Tommy Steele died, his Ford V8 cattle truck was taken over together with its driver Jimmy Churcher. Joe Page at Denmead sold his two vehicles - a Bedford and a Vulcan - but when another vehicle was bought from Jobsons of Hambledon Road, Denmead, the Jobson house and yard were also purchased. This was to be the Hill Denmead base for several milk vehicles working that area.

At the outbreak of World War II, eight vehicles were in service. One of the regular contracts at the time was to haul churned milk from the Isle of Wight Newport Creamery to Portsmouth. When the creamery was bombed, Hills lost one of their vehicles although fortunately the driver was unhurt.

The Hill family were to double their operation in 1944 when Leonard Percy sold his Meonstoke milk vehicle operation although the end of the War was to see a big change in the Hill business structure - and the sort of the jobs they would undertake.

During the early 1920s, VA operated a Model T Ford to take his own milk to the Portsmouth dairies but the work grew when other farmers asked him to do their milk transport as well. The Ford was replaced by this Morris Commercial van, which was usually driven by VA's eldest son EWG. Eric is seen on the left, with his father, outside Hillview, the farm of Simon Horn. The link between the Hills and the Horns grew stronger after Eric married Simon's daughter Maud in 1932.

The photograph may be fuzzy but this Dennis 30cwt four-wheeler - PX 7107 - marked an important milestone in the haulage destiny of Hills. Their first proper lorry was bought second hand in 1928 from Sparshatts Garage at Hillsea, Portsmouth (pictured opposite). EWG was the Dennis' first regular driver and he was expected to handle the ex railway 17 gallon milk churns then in use. Eric collected from the Swanmore, Bishops Waltham and Droxford areas. Once the milk round was done, the Dennis was used for all manner of farm related traffic. This included many loads of hay and straw to the Pickfords and Carter Patterson's stables near to Portsmouth railway station.

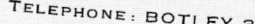

In 1932, VA moved his family to the Bottings Hotel opposite Botley railway station. The Hotel took its name from the previous tenant - Harry Botting - and as well as taking in guests, it did all manner of other trade and VA & his sons continued this. On behalf of the Southern Railway, parcels were delivered to Botley, Hedge End, Curdridge, Shedfield and Shirrell Heath areas. One of the Hotel's two taxis was the Wolseley, which the pictured (centre) George Trivett usually drove. George did the longer runs with parcels although Joe Blake - and the hotel horse Laddy - delivered the local deliveries. These included kegs of fish to the Ince fish & chip shop in Botley and to the similar outlet ran by Mr & Mrs Cummins at Hedge End. Joe did all sorts of odd jobs around the hotel and gardens and is seen in the coachman's seat (bottom left) with one of the taxi drivers (and a friend) and also with the Hotel's milking cow Molly (bottom right).

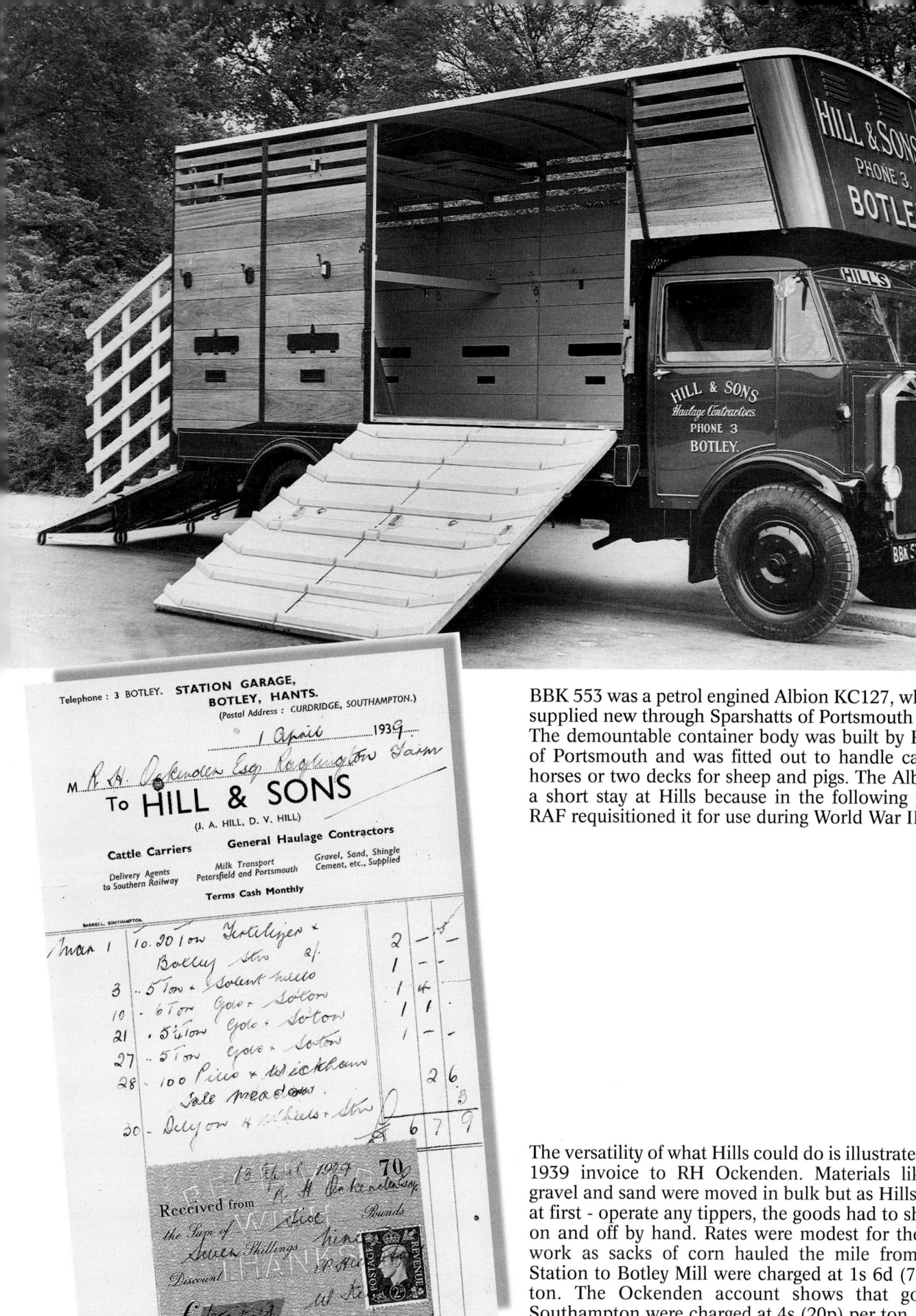

BBK 553 was a petrol engined Albion KC127, which was supplied new through Sparshatts of Portsmouth in 1938. The demountable container body was built by Readings of Portsmouth and was fitted out to handle cattle and horses or two decks for sheep and pigs. The Albion had a short stay at Hills because in the following year the RAF requisitioned it for use during World War II.

The versatility of what Hills could do is illustrated in this 1939 invoice to RH Ockenden. Materials like coal, gravel and sand were moved in bulk but as Hills didn't - at first - operate any tippers, the goods had to shovelled on and off by hand. Rates were modest for their local work as sacks of corn hauled the mile from Botley Station to Botley Mill were charged at 1s 6d (7.5p) per ton. The Ockenden account shows that goods ex Southampton were charged at 4s (20p) per ton.

The Hill business enjoyed a major expansion when Len Percy's firm at Meonstoke was bought about 1944. As well as the six vehicles, the premises (shown above) were also acquired as an operating base. Len Percy is seen (centre) with an early parcel delivery van although he soon got into milk haulage, again using Fords. Johnnie Adams is the pictured driver and he was to marry Len Percy's niece Gladys. When the Adams' were expecting their first child, Johnnie approached Jack Hill for a job. Hills were paying £2-5s per week to their milk drivers rather than the £2 that Percys were paying, although it cost him an extra 2s 6d on rent (when Johnnie moved house). Johnnie stayed at Hills for a long while and only changed employers when Hills sold their cattle wagons in the 1960s to Witchards of Alresford.

Bill Lashley is seen with his 1930 Ford 25cwt churn carrier, which he drove for Percys. The milk churns were known as Railway type and could carry 17 gallons. Getting the loaded ones off the back of your vehicle (if you couldn't transfer straight across onto a loading dock) required a special technique but most of the paving stones outside the dairies were all cracked due to them being dropped down the driver's leg.

As Percys became better established the Fords were replaced by examples of the Dennis Ace. Second hand Vulcans were also bought during the wartime as the military and essential users always had the first option for any new vehicles, which were made.

DISPOSALS AND ACQUSITIONS

Unless you were around at the time, it's hard to imagine what things were like during the early 1940s. While the main priority was to survive and win through the War, life did go on. General haulage operations were co-ordinated under the umbrella of the Ministry of War Transport (MOWT) although a variety of other Ministries were also involved in the movement of food produce including the Hill staple traffic of milk and livestock.

As a local Transport Group Controller, Jack Hill oversaw about 2-300 vehicles. Hills Transport was registered as an essential work establishment and their traffic also included all manner of wartime munitions. Their involvement included being on call 24 hours a day for whatever emergency might arise. One call during the night required diving gear to be rushed from Portsmouth docks to Lands End when a submarine failed to surface. As the invasion force to re-take France was being formed, the Hill proximity to the English Channel meant the intensity of work grew as D-Day approached.

The variety of what was expected from the Hills put them in good stead for what was to come after the end of World War II. During the six years of conflict ('39-'45) various depots and stores had been built up all over the country so once hostilities ended, these materials had to be disposed of.

Hills were engaged by the Ministry of Supply to disperse and dispose of the stores of eight different depots. These included the Naval depot at Exeter and three in Southampton while the two largest were at Newbury racecourse (which housed 300,000 tons of surplus equipment) and at Taunton.

Liasing with auctioneers and handling a variety of equipment meant that Hills were employing up to 500 people on each site. The main part of this work was to last from 1946 to 1948 and the experience involved was to have a huge impact on the destiny of the Hill family.

In 1946, the transport company of Hill & Sons (Botley & Denmead) Ltd was formed with Jack and Dudley as its two directors. Their father VA moved from Bottings Hotel to start up market gardening again at Chapel Road, Curdridge. Their elder brother EWG also decided to go back into farming at Snakemore Farm in Durley although due to a spin off from the MoS disposal work, Jack and Dudley were to also head off in different directions.

As part of the depot clearance work, a number of temporary Army huts had to be moved. Dudley appreciated there was a potential market for buildings like Nissen huts and what started on a very small scale, was to eventually spawn the company of Hill Construction Co or 'Hillspan' as it was often known. The huge difference between road transport and selling buildings forced the brothers to decide on going in two different directions. While Dudley took on Hillspan (see page 124) Jack put all his efforts into building up his transport business - at the same time as he endured the pressures of the late 1940s Labour Government plans to nationalise road haulage.

The creation of British Road Services was a massive undertaking although even at the outset a number of traffics were specifically excluded. The regular Hill work of milk and cattle haulage was amongst those, which weren't touched. Round timber hauliers were also exempt from compulsory nationalisation so Jack Hill was able to buy the specialised timber haulier of H Webb & Sons (Marchwood) Ltd and utilise their expertise to develop heavy haulage work - another exempted traffic. Of course once the Labour Government was unseated and the Conservatives re-elected, Hills were soon in line to buy up some ex BRS vehicles and get back into long distance general haulage. The 1950s was set to be an exciting time.

ROAD HAULAGE ASSOCIATION
LIMITED.

This is to Certify that

.... Messrs. Hill & Sons, ...

OF ..... Botting's Hotel, Curbridge, Southampton, Hampshire, ...

WAS THIS DAY ELECTED A MEMBER OF

ROAD HAULAGE ASSOCIATION

Dated this ... 1st ... day of ... January ... 1945

Chairman.

Member of the Council.

Director.

Registered No. 10546.

ARTICLE 13 (e) PROVIDES THAT :—

"*If the Member's subscription shall be in arrear and unpaid for three months after the same shall have become due and a resolution for the removal of such Member shall have been passed by the National Council after consultation with the Area Committee of the Area in which such Member is a Member. Any Member in respect of whom any such resolution is passed shall ipso facto and immediately cease to be a Member and shall not be entitled to claim a return of any money paid by such Member to the Association by way of subscription call or donation.*"

This Certificate is the property of the Road Haulage Association Limited and must be surrendered to the Association on cessation of Membership.

Although Hills joined the RHA in 1945, this was simply a continuance of membership of the forerunner to the RHA - The Associated Road Operators - which Hills joined in the 1930s.

HILL & SONS
HILL & SONS

HILL & SONS
HILL & SONS
071 YB
HILL & SONS
CTP 336

A major undertaking for Jack and Dudley Hill was when they were engaged by the Ministry of Supply, to disperse and dispose of the accumulated stores from eight different wartime depots spread across the south and south west of England. The one at Newbury racecourse alone housed some 300,000 tons of supplies so the project required a huge amount of manpower, with a variety of expertise to handle the range of materials. The main part of this work lasted from 1946 until 1948. It was the disposal of temporary buildings like Nissen huts (bottom right) which created the idea which formed the Hill Construction Co or Hillspan, as it became known.

Timber haulage work was specifically excluded from those undertakings compulsorily acquired under the late 1940s Labour Governments phase of road transport nationalisation. Hills were able to extend their operations by buying up the shares of H Webb & Sons (Marchwood) Ltd, a long established specialist in timber work. Herbert's son is seen in March 1931 driving the Fordson tractor with a load out of Ashbridge Park destined for Marchwood. Not a lot of weight is being hauled by the AEC Matador in November 1950, which took this 70' flagstaff, made from an 80-year-old New Forest larch to London as part of the 1951 Festival of Britain festivities. In developing this aspect of work (where they tended to favour winch equipped Foden tractors) Hills kept three of Webb's specialist pole trailers and a quartet of their AEC Matadors. The remainder of the Webb equipment went under the auctioneer's hammer on 17th August 1951. The vehicles offered for sale then included a 1943 Dennis 5 tonner CTP 23; a 1946 Fordson DTR 964; a 1939 Leyland Lynx CYS 792; a 1936 Leyland Cub CWJ 330 and two Bedford 5 tonners - ECG 475 and CTP 306. Also listed in the sale catalogue were the AEC Matadors: FOU 361; GOR 206; GLU 866 and a six-wheeled version GAA 311.

With the farming community making up the largest slice of the Hill clientele, naturally the haulier exhibited at most of the area's large annual agricultural shows. The lower shot shows the Hill stand at the Royal Counties Show at Reading in 1953. Five years earlier, Hill's brand new Dennis Horla (pictured above) had made a big impression at the 1948 Salisbury Show. The outfit's semi-trailer sported the latest state of the art Vincent bodywork that could accommodate six horses plus a special compartment for the grooms above the tractor's fifth wheel.

JRD 149 was also a specially built horsebox, the Bedford chassis supporting Jennings of Reading bodywork that could carry three horses. This Bedford usually had Len Samways as driver and it replaced a smaller Bedford that had been bought when Hills took over the vehicles and 'B' licences of Fred Seaward. In the main these were cattle trucks although Hills soon replaced one with a Bedford tractor unit coupled to a Taskers Little Giant low loader to move plant around.

By 1952, Hills had 17 vehicles dedicated to daily collection of milk - 7 days a week. Three more vehicles came into use during the Spring time flush when cows produced extra milk. Unlike today where milk is generally collected (in bulk) direct from the farmer's yard dairy, in this era churn pickup was also done from custom-built stands, normally at the farmer's lane end. Titch Emmett is seen loading his 1945 Thornycroft on the Preshaw Estate north of Bishop's Waltham. Hills were very keen to enjoy the early benefits of articulation and even used artics on collection work. In 1951, Hills collected 4.5 million gallons of milk and travelled 335,000 miles in the process.

Jack Hill quickly identified a need for milk tanker operation although their loads were generally the waste products of skimmed milk & whey. This was sold on to local farmers, as it again proved ideal for animal feed. The tanker division of Hills also carried molasses in bulk and fuel oils.

Seen from the air, the huge differences in the potential of the two Hill depots of Botley (top) and Denmead (below) are apparent. The railway line and Botley station are in the right area of foreground with the Bottings Hotel still standing proud. New buildings are in the process of being erected but it was the huge area of orchards and fields to the top of shot which proved invaluable after the Hills operation dramatically expanded.

Although Hills varied traffic meant they weren't part of the late 1940's nationalisation programme, Jack Hill was keen to get back into general haulage as soon as British Road Services began selling off some of their fleet. He's pictured in April 1954 shaking hands with BRS Manager Thornton at the hand over of two AECs. Jack's son Chris is enjoying the moment while Hill's workshop manager Bert Joslin is also pleased to be behind the wheel.

In total, Hills took delivery of 34 vehicles from BRS and although some were in scrap condition, it was the vehicle's Special 'A' carriers licences (which allowed a haulier to carry anything, anywhere) that was the main part of the deal. Hills took the first vehicle to be sold out of Southampton depot and with this Thornycroft are (left to right) Bert Joslin, Bill Ponsford, Jack Hill and BRS Manager Thornton. The similar 1951 Thornycroft ZE/TR6 four wheeler JCR 691 was bought as part of BRS lot: 66/361. Jack Hill paid £3,100 for the pair.

With the early 1950s programme of partial denationalisation of BRS, Hills return to general haulage was done with a variety of vehicles. Hills had a contract for the storage & distribution of Pascals sweets and the Reading based AEC Mammoth Major (top) - new on 1st April 1942 - is seen about to deliver another 15 ton load. FBK 20 is a diesel engined Albion CX1 (new on 11.3.50) and was one of two vehicles bought by Hills (for £2,525) in the BRS lot no. 66/670 when sold from the Southsea depot. Hills operated it at Reading. The CX5 Albion six wheeler - also of 1950 vintage - is loaded with market garden produce. It was rated to carry 12/13 tons and had a 105bhp 6-cylinder Albion diesel engine.

Hills could turn their hands to almost anything. The 1955 S type Bedford (above) is receiving Guernsey flowers, which have just been flown over for the Southampton market. The Leyland Beaver (below) was based at Reading and used for grocery deliveries. VME 693 was chassis number 494448 and while built in 1949, was first registered on 30.3.50 to S Smith & Sons of Edgware Road in London. Originally petrol engined, the first year's tax disc cost £85. In 1956 it was fitted with a diesel engine and by that year, the annual tax disc cost £105. An extra £20.18s.6d of revenue was paid so that it could pull a drawbar trailer. The Hill Group bought it in June 1959 but when entered into the big Hill auction of 1967, it didn't raise a bid.

It was part of the Hill attraction - so far as the independent transport observer was concerned that Hills of Botley had their own way of doing anything. The 1952 S type Bedford artic (below) with a standard size of Scammell semi-trailer was always going to be a general haulage outfit although Hills still expected it to carry a variety of different loads when required.

Meanwhile, the 1947 DG Foden (above) was primarily intended for heavy haulage or timberwork but once it was hooked up to a suitable dolly converter, it could also turn its hand to general haulage traffic. And of course an eight wheeled Leyland Octopus and trailer (centre) can carry all sorts.

One of the joys of being involved in heavy haulage is you're expected to move all manner of different loads. When you drove for Hills you were expected to do the job with all manner of motors - in varying conditions. The ex Webb AEC Matador (centre) is hauling a 4x4 Federal F55 aircraft towing tractor. With a 9'9" wheelbase, the Federal had all wheel steering and tipped the scales at 33,000lbs.

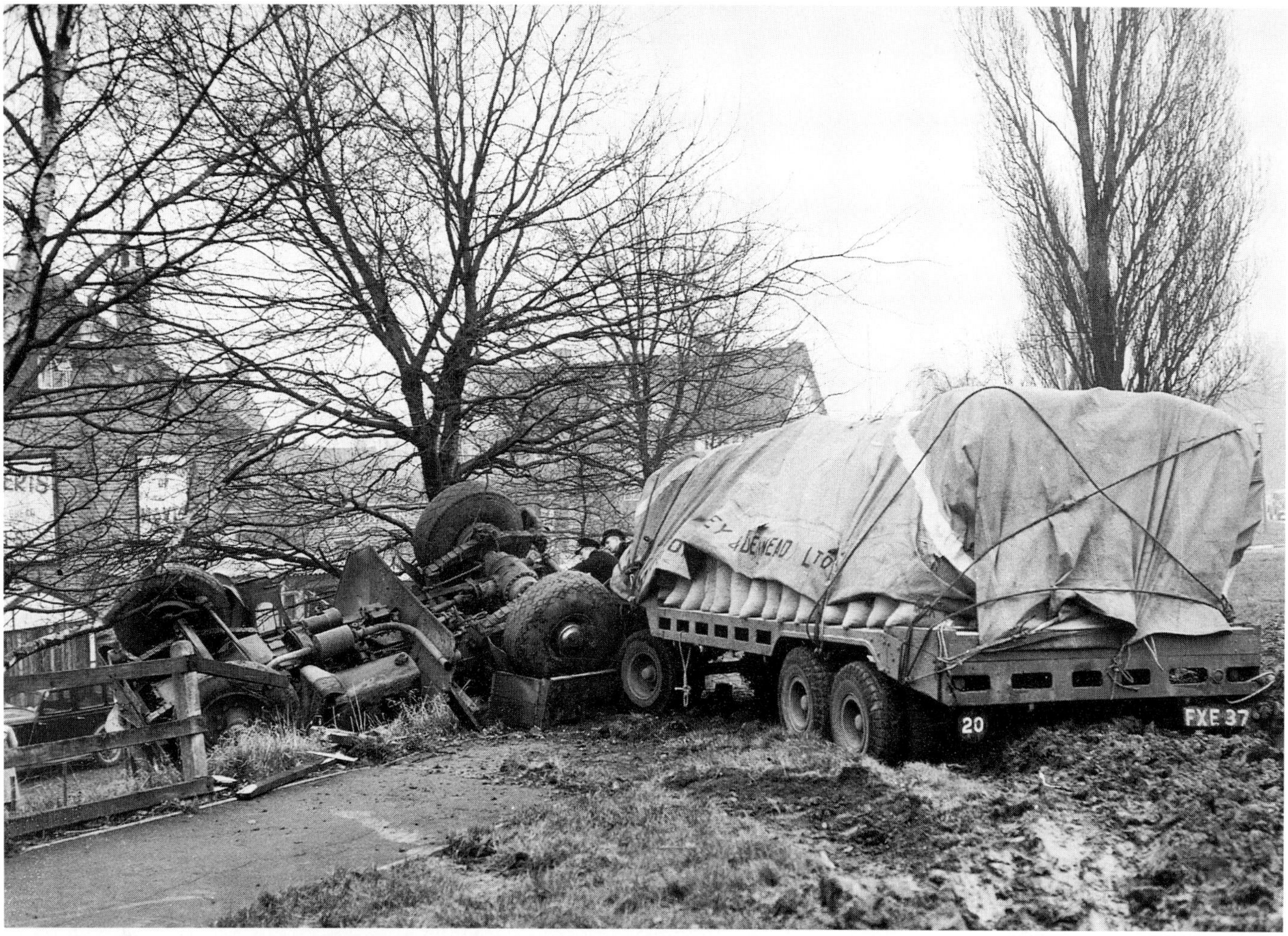
20
FXE 37

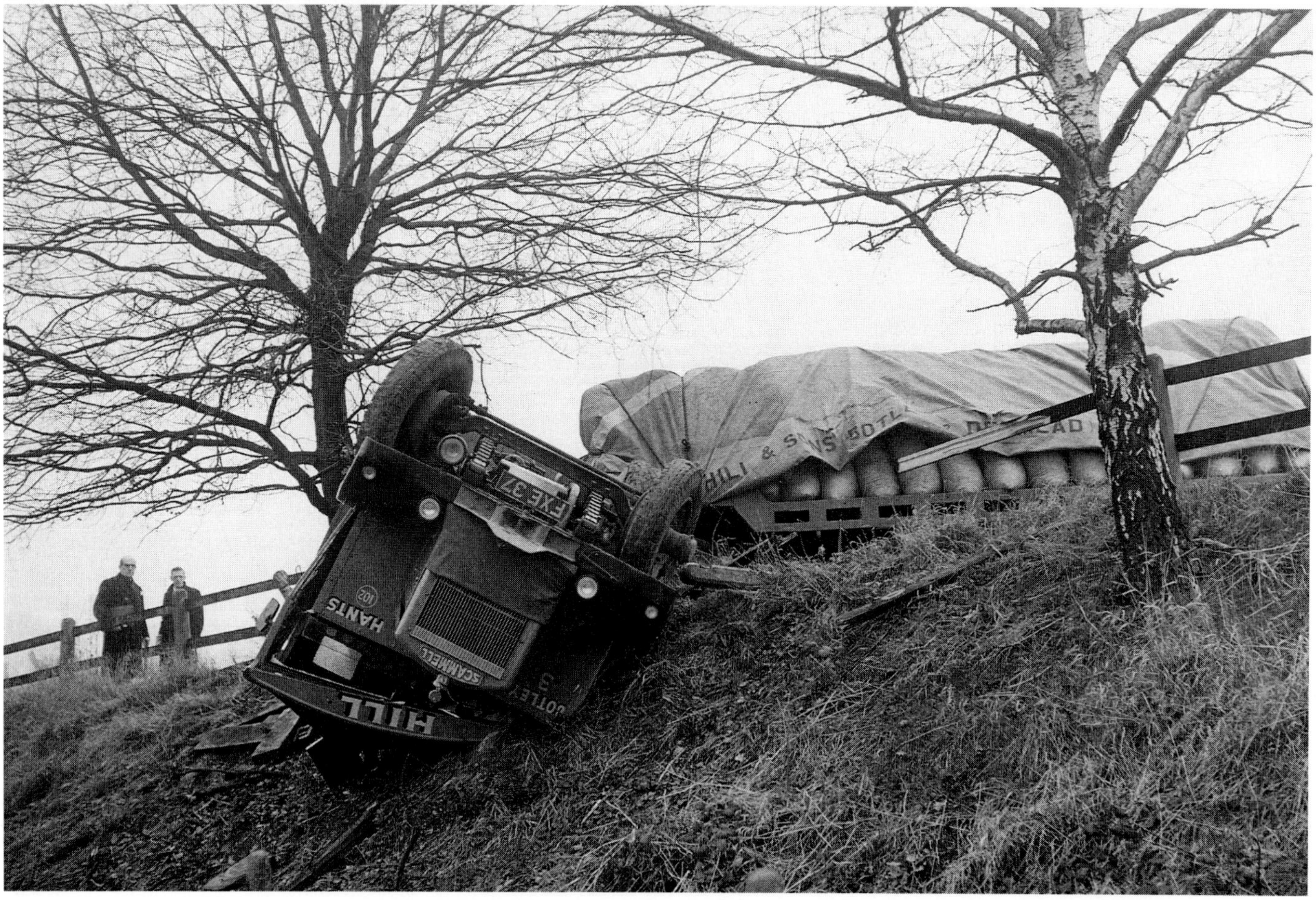
FXE 37
HANTS
SCAMMELL
HILL

When you're operating a fleet the size of Hills, it's not surprising that accidents do happen. An errant car driver who forced the Hill vehicle to take avoiding action caused this one. The 1939 Scammell was loaded with 13 tons of cattle feed when it left the road at Hinchley Wood in Surrey. Kenneth Sessions was driving and the only injury he sustained was getting acid burns from the Scammell's battery. He agreed the tree had been handily placed to stop the artic going any further and even the Scammell was repaired. FXE 37 did all sorts of work for Hills but by the time it reached the big Hill auction of 1967, it was without an engine and it didn't even raise a bid. It had started life with Shell Mex as an artic tractor unit although Hills converted it into ballast tractor form.

Jack Hill had been a convert to the joys of articulation from as early as the late 1940s. HGT 452 started life in 1947 with one of the large petrol companies and is seen hauling an old railway carriage which Jack Hill's elder son BJ converted into a birdhouse - although parked it in the depot. The DG Foden (centre) is loaded with what Hills were told was Montgomery's office during the World War II. It was bought and transported to the Tower of London car park where it became Hill's London office. Charlie Tigwell usually drove this Foden low loader. The Mark 5 Seddon HCR 207 (bottom) dates from 1950 and has driver Walt Jacobs leaning against the rear wheel. While both semi-trailers in this shot were known as a 'Queen Mary' and built to carry aeroplane parts, the foremost was an awkward 9' wide compared to the 7'6" version hauled by the rearmost Albion tractor unit.

Fresh out the Hill paint shop (although it was first registered in 1950) the 6x4 Scammell Pioneer KXT 872 has driver Gordon Prebble at the wheel and 45 tons of Thornycroft boiler on the Eagle drawbar trailer. Seen in Titchfield during early November 1959, the 20' high outfit took 10 hours to travel from Woolston, Southampton to the Admiralty Fuel Experimental station at Haslar with overhead wires and low hanging tree branches being the biggest headache. Charlie Knight is believed to be the figure on top of the load.

TRAINS & BOATS & PLANES & TANKS

Hills moved a large amount of rail rolling stock and were happy to be involved in moving the same engine twice - albeit with a gap of 13 years in-between. The 'Hayling Billy' operated solely between Havant and Hayling Island from 1928 until 1964 but when the service was discontinued, Portsmouth Brewers bought the engine to use as a sign on the forecourt of their new 'Hayling Billy' public house at East Stoke on Hayling Island. Charlie Tigwell and the 1956 Scammell box tractor MUH 21 did this move in 1966 using a Crane float trailer (bottom). 13 years later, the Whitbread Wessex Brewers (who had taken over the 'Hayling Billy' pub) decided to donate the 102-year-old Terrier class engine to the Isle of Wight railway preservation group. Before the engine was fully restored, the Wight Locomotive Society re-painted it into its original olive green livery, renumbered it and renamed it with its original proper name of 'Freshwater.' Jack Hill is on the engine's footplate while elder brother Eric is beside the trailer. Eric (EWG) had come back to work at Hills after retiring from farming. He first ran the Bridge filling station before taking charge of the company workshops.

With such a close proximity to the sea, it's not surprising Hills found themselves involved in moving a huge variety of boats. Awkward to load, awkward to carry and generally vulnerable to being damaged, boats usually required sympathetic carriage using the right cradles and securing methods. One thing these loads all shared whether it was the small yacht moved by the Mark 5 Seddon in 1958 (top) or the more substantial launches seen in Botley square about 1956 was they were tremendous head turners.

Thornycroft produced all manner of top quality boats and Hills of Botley were able to offer their expertise when they had to be moved on dry land. One of the most awkward was the pictured diving craft (centre & bottom), which was moved from Southampton to Tilbury docks prior to onward shipment to America. Based on a Rogers tank transporter, Hills designed this particular boat trailer, which had a 12' width at the rear. With an 18' running height, the hauling tractor was Hills much travelled Scammell Explorer TOT 297, which not surprisingly had the company nickname of 'Tiny Tot.'

The hardest part for Hills in moving this Bournecraft built motor vessel was simply extracting it from the works. Although pictured being sheeted up, it had required the efforts of two of Hills specialist box tractors to get the supporting trailer round the building on the right hand side. While one tractor pulled from the front, a second tractor pulled the trailer sideways - on top of some greased sheets. As seen in the colour section (page 53) the road haul to London was done using a Scammell Junior Constructor.

As the proud motif on the side of this boat suggests, this World War I motor torpedo boat gained the Victoria Cross. Hills moved the boat from Gosport using one of their three Albion Clydesdales but made this stop in Alton outside the house of the then boat's captain.

Hills have handled all sorts of vessels although this internal move in Vosper Thornycroft's yard from the prefabrication shed to the outfitting quay during 1966 was certainly exacting. Simmons Engineering converted the Rogers tank transporter trailer to support massive boats like this. At 30' long and 18' wide, the vessel is being pushed by the Scammell box tractor MUH 21. The boat was one of two Rolls Royce powered patrol boats built for the Kuwait Government.

Having one plane crash on landing suggests an accident but when two similar Hawker-Siddeley 748s belonging to Channel Airways both crash landed at Portsmouth Airport within 90 minutes of each other, a major enquiry ensued. The incidents occurred on Tuesday 15th August 1967 and generally speaking it was the waterlogged state of the runway, which caused the pilots to loose control on landing. Both planes regularly flew in and out of Portsmouth and perhaps the one plus point to the incident was the second one ended up - in stricken condition - at a different point to the first one albeit through the perimeter fence and blocking the main Eastern Road for five hours. The first 748 had been enroute from Southend to Paris (via Portsmouth) while the second plane was returning from a regular trip to the Channel Islands. In total 70 passengers and crew were onboard the two planes but no one was apparently hurt. Hills of Botley helped to pick up the pieces by moving the damaged planes across the airport into a suitable hanger. Although two different tractors were used (the 1952 Scammell XMT 45 and a '65 Albion AOR 325C) for the two different planes, Bob Bernard drove them both.

Some aeroplanes may be lighter than some boats but they're certainly just as awkward to handle. Driver Murphy - on the right - and mate 'Macca' McAlpine (on the left, both top photograph) take time out near Blackbushe aerodrome on their two-day trip from De Havilland's Christchurch factory to their premises at Hatfield during August 1961. The flying wing had a travelling width of 22'. Charlie Tigwell is driving XNM 477 (bottom) carrying an Avro Argosy fuselage, which was 75' long, 11'6" wide and 15' high. The Scammell Highwayman is coupled to Hills home made Hovercraft trailer which incorporated steering on its two axles.

Moving battle tanks - and similar military crawler driven vehicles - has been a Hill speciality for the last 50 years. Ex military trailers, mainly of American build, proved versatile enough in standard form although Hills preferred to modify them for all sorts of other uses. Seen unloading a Chieftain at Tilbury docks (top) driver Bill Miles is the figure on the left. The Rogers tank-transporting trailer has had a different pair of axles fitted under the neck to increase capacity. There's about 52 tons in the Centurion Armoured Recovery Vehicle (centre). These were built between 1948-58.

The 50-ton Dyson tank transporter incorporated a sloped build to allow for easier unloading but also to allow for the front steered axles to turn.

Hills KXT 872 - and the Scammell Junior Constructor of Hallett Silbermann - are both carrying Flail tanks mounted on Churchill chassis'. Used for mine clearance work, these tipped the scales around the 45 tons mark.

There's about 30 tons in this Sherman tank, which is supported on a Crane Mk I trailer. Hills chopped down a number of this style of trailer to use as bogies.

This amphibian was moved from the Military tank park at Bovington to Ringwood during 1960. The tracks acted as paddles when the vehicle was in the water but their narrowness meant they were awkward for Hills to transport as they could easily slip off the side of the road going trailer.

48

The first Hill vehicles were painted an austere looking brown. Jack Hill was a keen exponent of articulation and the concept gave him the opportunity to squeeze a bit more weight onto his vehicles. The heavyweight road roller seems an ideal fit on the Taskers Little Giant semi-trailer but the stance of the Bedford tractor unit indicates it has its fair share of weight onboard.

When Hills bought out the Webb of Marchwood business in 1951, one of the ex RAF AEC Matadors, which Hills kept (another four were sold in the Webb clearance auction of August 1951) was GHO 29. Webbs first registered this 4x4 on 13th February 1948. Retired driver Jim Purkis recalls the Matadors were expected to move weights of over 30 tons and although they were good enough performers, the lack of trailer braking was his main concern. When the scrap dealer Harry Pound bought up a large number of tanks at Woolwich, Hills had the job of taking them south to Portsmouth for scrapping. Paul Hancox identifies the load as probably an unarmoured LVT 4 (Landing Vehicle Tracked). More than 8,000 of these were built between 1943-45 and the British Army called them 'Water Buffalo'.

49

The marque of Seddon was well used by Hills in a variety of shapes of models. Not a bad performer said one of their old drivers, provided it was fitted with an Eaton two-speed axle. The colour of XXW 534 isn't a trick of the light but a reminder that this tractor unit had just been bought from Whitbread Breweries. The motif on the driver's door incorporated a map of the British Isles as an indicator of the area which Hills served with their fleet.

Hills only operated a small number of AEC Mercurys mainly out of their Reading depot. This Mark I was first registered in 1954 at Warrington. Seen during May 1963 with an awkward crane jib load, this four-wheeler had an unladen weight of 4 tons 16cwt. It was originally designed for 12 tons gross operation - until it came to Botley.

PAA 805 was bought new on 16th February 1956 and was rated in artic form as having a 10/12-ton capacity, which was hard work for a small Albion Chieftain. It is seen in May 1960 about to leave Industrial Pressure Vessels at Eastleigh. Based at Hills' London depot, it ran with a Special 'A' licence allowing it to haul any goods, anywhere. The Alford brothers regularly drove two of Hills Albion artics.

The Scammell 25 tonner HGT 452 was new on 13th March 1947 although Hills of Botley didn't buy it until November 1958. It was operated on Hills 'B' licence (with Charlie Tigwell as its first regular driver) with an unladen weight of 8 tons 9 cwt purely as a replacement vehicle and could only be used in the event of another licensed vehicle breaking down or undergoing repair. Hills moved a lot of material similar to this six-wheeled American La France Type 0-11A crash tender (with bodywork by Foamite) for the US Air Force pictured in March 1963. The freight was shipped in through Southampton docks and bound for bases like Greenham Common and those in Bedfordshire, Cambridgeshire and Northamptonshire.

Driver Charlie Tigwell was rewarded with a new motor in October 1961 when XNM 477 arrived at Botley. Bought through Rush Green Motors of Hitchin, the two-year old Scammell Highwayman was new to HG Pentus Brown of Leighton Buzzard. Pictured in July '63, the Gardner 150 powered tractor unit was nominally rated as a 25 tonner and had a licensing weight of 9 tons 6 cwt.

The ex Marples Scammell Junior Constructor 874 AUU did a huge variety of work for Hills in the 14 years it was at Botley. It's pictured (top) en route to the 1965 Earls Court Boat Show with the event's largest exhibit "Solaria". Built by F Bourner Coach Works Ltd of Hamble, the vessel was so big that it was left on Hills trailer for the entire event. Moving this size of excavator (bottom) on a four axled girder outfit meant tyres could often be a problem and it was such a failure that prompted this time out. Barrie Hill is amongst the crew of five taking this machine to Bridgewater.

I have no qualms about re-showing a shot of 358 ETN, which is also in the book: '100 Years of Heavy Haulage'. The Scammell Constructor is pictured in June 1964 with a concrete beam destined for the new Havant by pass. With an unladen (licensing) weight of 14 tons 14 cwt, the big winch equipped 6x6 (new in 1958) did all sorts of work for Hills. One thing you can never accuse Jack Hill of is not trying to get value for money from whatever vehicle he bought. Even though this tractor eventually reached a point of being of no further operational use to Hills, he wouldn't part with it - unless some one would pay for it. He even tried to persuade me to buy it in 1986 when it was fighting a loosing battle with the local blackberry bushes (bottom). It was eventually sold and it was last heard (in '98) of being in Bedfordshire.

TOT 297 was another Scammell, which gave its Botley owners extended service. First registered in 1958, the ex military 6x6 Explorer had its original petrol engine replaced with a 680 Leyland diesel. It too is seen with a beam destined for Havant by pass (top) and also hauling girders into Fawley power station (bottom) in January 1962. The latter job gave Hills many months of work as it involved moving something like 40,000 tons of constructional steelwork with girders up to 90' long. The job of unloading the rail trucks at Hythe goods yard and delivering the steel four or five miles to site originally went to Pickfords. However, the mobile cranes, which Pickfords sent to the job, couldn't lift the heaviest beams so British Rail asked Hills if they could take over. Although Jack Hill didn't have any heavyweight craneage, he used a pair of Foden timber tractors to winch the heavy beams straight from the rail trucks onto the heavyweight road bogies. Jack is one of the two figures alongside his Mark X Jaguar saloon car.

The 1954 registered Leyland Octopus OJJ 700 came to Botley around 1962 but only stayed for a couple of years before all the cattle vehicles were sold to Witchards. It made its mark when involved in moving the Chipperfield elephant (see pages 92 & 93). Generally the Hill cattle vehicles operated with demountable 'float' boxes, as the weight of the box didn't count in the unladen weight - critical for both taxation and carriers licensing purposes. This factor also helped two Hill drivers to get off charges of speeding in 1954. Without the boxes, the vehicles were under 3 tons unladen and could travel at 30mph. The police had originally calculated their weight - with the boxes - as being over 3 tons thus restricting the vehicles to the lower limit of 20mph.

During the 1960s, Hills changed their fleet colours from a deep maroon to light blue. Apparently the change was prompted after a second hand vehicle in that colour was bought and because it looked so good, the change was decided upon. The new colours prompted this mid '60s line up of some of the various artic units then in service - three Albion Clydesdales, two Dodges, two Seddons and an AEC Mercury. The figures are Chris Hill (on the right) with his Uncle Eric (EWG).

It may look like a Guy and was actually made with Guy parts but this 6x2 tractor was built at Bridge Motors and proudly called a Hill. It was the first of three similar Invincible looking units plus a fourth which resembled the Guy Big J.

Hills of Botley vehicles were a regular sight around Hampshire but the feats they performed, often made people stop and stare. Loads didn't come much more impressive than the two columns taken into the Esso Oil Refinery at Fawley on Sunday June 28th, 1970. As shown in detail on pages 108 and 109, the route from the Terminus Station sidings at Southampton via Western Esplanade, Four Post Hill, Totton by-pass and then Marchwood by-pass was as testing as it came. The two vessels were both about 150' long overall and weighed in at 53 and 39 tons. Amongst those involved on the job were Chris Hill (steering the ex WD tank transporter supporting SPT 600's load), Dave Levy and Roy Wiseman. The two drivers were Bill Miles and George Holbrook who took seven hours for the trip although regular stops to ease the flow of traffic was the main reason behind the lengthy journey time. What made the job that much sweeter was the remark in the letter from the vessel manufacturers - Cammell Laird (see page 107) - that one competitor apparently thought the road move was impossible.

SJD 801F was new to Pickfords in January 1968 but was bought by Hills of Botley (via the Ewell based dealer George Hardwick) in January 1980 for £5,000 + VAT. Pickfords operated this 125 ton rated Scammell Contractor in ballast tractor form (although Hills later converted it into an artic) and it's seen in Southampton docks not long after purchase. Driver Peter Howlett has just hauled this American built locomotive off the ACL ferry from America. It was the first of four similar rail engines destined for shunting duties with the large quarry concern Foster Yeoman.

The Scammell S26 HBM 529Y and the S24 A390 OPX (plus the similar S24 LTR 689Y) were to be the Hill flagships throughout the mid and late 1980s. Both HBM and A390 arrived at Botley on the same day - 30th July 1983 - and while the 150 ton gross S24 wasn't registered until 5th August '83, Scammell Motors had actually used it for prototype appraisal of the fairly new S24 range. The similar 6x4 S26 (rated as an 80 tonner) cost the same amount as the S24 A390 (£24,000 + VAT) but it was bought almost new from Yardley Commercials. Both the Scammells are seen in Southampton docks with machines destined for Redditch. Dick Lodge was driving the S26 and it's hauling a 35-ton Komatsu machine. Keith Knight is seen with the S24 coupled to a 1977 tri-axle Walker low loader carrying a 45-ton Komatsu excavator.

Easing their way through overhead foliage was a task, which the Hills of Botley crews regularly performed (top). Pictured on the A27 east of Southampton, this 26-ton boat was 68' long, 14' wide and 18'6" high. The craft is being moved from Comer Engineering in Botley to Moody's boatyard in Bursledon. Frank Hackyard was at the wheel of the Scammell S26 while his boss Chris Hill walked the entire route alongside the King four-axle, semi-low loader trailer. No foliage to worry about in Esso's Fawley refinery (right) with the S24 seen coupled to an extendable four axle Scheuerle that is supporting a 110' long column.

The huge size of the trawler 'Kittiwake' meant it was never going to be made far from the water. The builders used wasteland inside Southampton docks close to no. 7 dry dock to build the 60' long, 17' wide, 24' high vessel, which tipped the scales at 75 tons (giving a gross weight of about 115 tons). The job on 8th October 1989 was fairly straightforward - albeit highly impressive - for driver Keith Knight, mate Roy Wiseman and Jim Purkis. After Grayston, White & Sparrow lifted the boat onto Hill's four axle Scheuerle semi-trailer, the crane was moved so it could then lift the boat off the Hill cradle once the S24 had taken it to 101 berth where it was named before going into the water.

G992 ROD was bought from the contracting brothers, John and Trevor Coles of Milton Daverel in Devon at the end of March 1990. The 6x4 Scania 143-450 was rated for 150 tons gross operation and at the time was only about 8 months old with only about 13,000kms on the clock. It gave good service to Hills during the 1990s (and was still apparently being worked in 2001 after being sold to a local operator). It's pictured when fairly new with a 2-bed-4 Nicolas semi-trailer but if you look closely, the tractor unit and trailer are not coupled together. The Nicolas trailer belonged to Foster Yeoman and was carrying a similar boiler to the one loaded on the Hills King trailer (parked behind) both being enroute from London to Southampton - the Scania was simply trying the trailer for size.

The road haul of this fully completed Halmatic vessel was limited to a mile in distance through Poole but the job still required a great deal of care and attention. A road was closed as the boat had to be craned over a building and then placed onto the four axle extendable Scheuerle modular semi-trailer. All up weight was about 106 tonnes with the vessel itself being 72 tonnes, 85' long, 16' wide and 29'6" high. G992 ROD (in four axle form) has Keith Knight at the wheel while Chris Hill is the figure in the yellow vest. In charge of the loading operation was Jim Purkis while Roy Wiseman was the trailer mate. The vessel was craned off into the water at the wharf adjoining the RNLI HQ.

Hills moved a large number of these overhead steel gantries which varied from 90-105' long and 20/30 tonnes in weight. They were taken from Waterman Offshore near Rochester in Kent to a variety of delivery points close to the M27/M3. Driver Keith Knight accompanied by mate Roy Wiseman are hidden behind the Scania's curtains although Ivor Truman and mate Vic Cooper appear ready for action. Ivor worked for Hills as a sub contractor almost exclusively for about 15 years. With the registration of 666, Ivor said that other drivers called his Cummins 400 powered Foden 80 tonner 'Sign of the Devil' but he recalls it as being a good motor. Roy Davy was another long-term sub contractor, he working exclusively for Hills for about 20 years.

The MAN 40.502 came second-hand to Hills during October 2000, it starting life with Dunkerleys of Yate in Avon. With all wheel drive, the 6x6 was something special in that it incorporated a torque converter in the transmission, which allowed it to be rated for 250 tonnes gross operation. It was no where near that sort of weight when Dave Weston photographed it on the Geest Banana dock in Southampton during August 2001 coupled to a five axle Nooteboom semi-trailer.

Driver Keith Knight and mate Roy Wiseman had their Scania 143 replaced with this specially built Iveco 470 150 tonner during May '99. The crew are seen in December 2000 with one of five module style loads (each weighing about 40-45 tonnes) which had been assembled close to the entrance of Calshot power station. At 19-21' wide and 22' high, the loads were taken at slow speed via the dedicated abnormal load gate into Fawley refinery.

HELPING OUT

Helping out was often the case in the heavy haulage game, especially if you were a long way from your home base. Wynns were some distance from their Newport HQ so appreciated the regular help of Hills. It's a sight to gladden the eyes of any AEC Matador follower. Being hooked up to the front of Pacific 192 'Dreadnought' must be prestigious in anyone's eyes. The Scammell FXE 37 also did its fair share of Wynns pulling and is also seen pushing (bottom) at Sheet, near Petersfield, while PC Ron Harris watches the action.

With a number of winch equipped tractors in their stable, Hills were regularly called out to extricate all manner of vehicles and machines. The efforts of Foden CTT 246 were required to assist this Pickfords outfit off the Red Funnel Isle of Wight ferry. Mr Pennington, Pickfords' Southampton manager is the dark suited observer.

This long serving tractor was re-cab'd and apparently re-engined before being downgraded to general dogsbody yet still had the strength to extricate this crawler tractor in trouble.

It's always great to receive a letter of commendation - especially from the local Constabulary - although dealing with an unexploded 500lb bomb seems well beyond the call of duty.

HAMPSHIRE & ISLE OF WIGHT CONSTABULARY

POLICE STATION,
FAREHAM.

Tel.No.:
FAREHAM 2285/6
Any reply to this letter should
be addressed to:-
THE CHIEF SUPERINTENDENT,
and the following reference
quoted: I.F./5833/59
Your Ref.

13th April, 1959.

J. Hill, Esq.,
Managing Director,
Messrs. Hill & Sons,
Station Hotel,
BOTLEY,
Hants.

Dear Sir,

My attention has been called to the ready assistance you have given the police on a number of occasions, including the use of recovery vehicles at the scenes of accidents etc., and, in particular, with the invaluable help in lifting a 500 lb. unexploded bomb at the side of the Bishop's Waltham - Botley Road on the 8th instant.

This kind assistance is much appreciated and I have much pleasure in offering you an expression of thanks from the police for your ready help.

Yours faithfully,

Chief Superintendent.

In approach and attitude, Frank Annis of Hayes in Middlesex was very similar to Jack Hill. Annis' were great believers in using ex wartime tank transporters for a variety of rolls and it's such a trailer being used to support this awkward ship's pro-peller. Hill's 'Tiny Tot' is giving a helping hand between Warminster and Shaftesbury in the early 1960s.

During March 1970, Wynns moved this transformer from Fleet to Portsmouth and with an all up weight around the 350 tons mark, the efforts of the ex Pickfords Constructor PUC 474 was needed to ease its passage. George Holbrook was the Hill driver involved.

Apart from a bruised ego or two there was very little damage after this recovery. The Naval Sentinel had backed partially into the River Meon - just off the A32 near Wickham - to recharge its water storage tank. But when it lost all its steam, it ran back into the river and couldn't extricate itself. It required the expertise of Hill's Scammell driver Bill Miles and mate Charlie Knight to restore everyone's dignity.

RATIONALISATION OF THE '60S

The late 1950s and early 1960s were busy times for Hills. The fleet expanded while new branches at Castle Street, Reading and Tower Hill in London (supplementing the original depots of Denmead, Petersfield, Meonstoke and HQ at Botley) reflected the spreading coverage of the Hills operation.

As far as the south coast area was concerned, Hills of Botley were something special. There may have been one or two independent hauliers in the area with larger fleets but no one did the variety of work that Hills did.

Their long-standing traffics included milk collection and cattle haulage but the company were specialists in such diverse things as heavy haulage and horse transport. They had custom built vehicles for meat deliveries, hauled liquids in bulk tankers while their long established products division bought and sold all manner of goods to and from the farming fraternity. They ran tippers, they did furniture removals and they'd even sold confectionery & groceries to shops round the Portsmouth area. Round timber haulage was a company speciality and they also ran a long distance, general haulage operation. If it were being moved by road then Hills of Botley would probably be doing it.

Hills were obviously one of the biggest employers in the area and a lot of the staff remained with the company for many years. Amongst the admin team was Mrs Ruth Harper who clocked up 30 years at Hills. She too can testify that Jack was certainly careful with his capital expenditure as up until 1995, she still used an old fashioned, plug in cable type switchboard to handle incoming phone calls.

The growth of Hills was not without its problems and after 30 years in the game, Jack Hill decided to take stock. As well as the day-to-day pressures, the diversity of the business brought a diversity of problems so in the early '60s, a plan of rationalisation was created which saw many of the Hill specialist sub divisions sold as going concerns.

Hills Reading depot was sold to Amey Transport at the end of 1963 while the cattle trucks were bought by EE Witchard Ltd. The horsebox vehicles were sold to Tarrants of Denmead and an agreement was reached with the Milk Marketing Board over the end of their collection commitment.

The late '60s Labour Government was again to have a huge effect on the structure of road transport. Barbara Castle is generally credited in being the driving force behind the 1968 Transport Act, which invoked all manner of legal requirements on the industry.

The need for HGV driving licences was re-created (the original legislation was allowed to lapse on the outbreak of World War II) while the 'O' Operators Licence replaced the controls then in force with 'A', 'B' and 'C' carriers licences. At one fell swoop, the value of a haulier's business was decimated because prior to '68, a vehicle with an 'A' licence - especially with the flexible conditions of: 'Anything, Anywhere' - was worth a huge amount. And with tight constraints governing the issue of any new 'A' licences, a haulier could make a tidy sum if he sold a vehicle with a licence attached.

In contrast, 'O' licences were initially issued to almost anyone who asked for one. The floodgates of people who could set up in general haulage for the price of simply buying a second hand truck put huge pressures of competition on established hauliers like Hills - never mind the likes of own account operators who prior to '68 could only carry their own goods. Replacing their 'C' licences with 'O's brought another whole section of competition.

Even more pressure was put on the industry with the requirement to plate and test every goods vehicle over one year old. The MoT requirement effecting cars over three years old to have their roadworthiness checked on an annual basis had never applied to HGVs - until the Transport Act was invoked.

This legislation had a big effect on Hills. Similar to many other big operators around the country, Jack Hill believed in getting value for money from his vehicles. He did buy some new vehicles but in the main, acquisitions came second hand so the fleet was best described as varied in makeup and age.

Having such a huge depot allowed Hills to keep vehicles which other people may have cut up and scrapped as they'd dropped out of service. Jack was of the belief that something may be of use later so why not push it up to the top of the yard and hang onto it.

However, things all changed in early August 1967 when more than 700 lots of what auctioneer John Jeffrey & Son described as: 'Hill & Sons extensive range of surplus equipment and surplus motors, tankers and trailers,' went under the hammer over two days of sales. The income eased the Hill financial problems but the clear out was to signal a big change in direction for the company.

Specialising in heavy haulage was to be the niche that Hills settled on and this expertise was soon taken on by Jack's son Chris, the next generation of Hill to head up the Botley concern.

As Hills had replaced their 1950 Scammell Pioneer's original Gardner 6LW engine (112bhp at best) with a 6LX-150, they expected it to perform all sorts of Herculean tasks. Jobs didn't get much more impressive than the five mile road haul on 20th March 1961 of two 53 ton barges (each 84' long, 20'8" wide and 13' high) from the Dellquay Hard beach in Chichester harbour to Portfield Gravel Pit on behalf of John Heaver Ltd (see also overleaf). The 56-wheel trailer was designed by Jack Hill and built especially for the job by Simmons Engineering, West End, Southampton. It was built from two 45 ton, 24 wheel Rogers tank transporters cut down and modified for the job. The first barge took two hours to travel the five miles although the second one only took half as long. Longest part of the move was the nine hours taken to chock up the barge on the beach (after it had floated in on the high tide) while steel plates had to be laid to build a short cut across the right hand turn at Chichester traffic lights. Gordon Prebble was the Scammell driver while Barrie Johnson was one of the many mates. In May 1963, Hills again moved these barges from the worked out pit to a fresh gravel area on the other side of the main road (see page 102).

PORTSEA ISLAND MUTUAL

Cooperative SOCIETY LIMITED

Reg. Office: 110 FRATTON ROAD

PORTSMOUTH

TELEPHONE 2221!

Date 29th January, 1963.

Our Ref.

Your Ref.

Please reply to DAIRY DEPARTMENT,
STATION ROAD, DRAYTON,
PORTSMOUTH.

TELEPHONE COSHAM 79422.

Messrs. Hill & Sons (Botley & Denmead) Ltd.,
Bottings Hotel,
Curdridge,
Botley,
Southampton.

Dear Sirs,

May we, during this lull in anxiety resulting from the severe weather which has caused so much disruption to the industry we serve, take this opportunity to express the appreciation of our Society, for the exceptionally arduous task of collecting milk from the farms, which was so ably overcome by yourselves and employees.

The heavy snowfall no doubt caught most people by complete surprise but the speed at which difficulties were defeated by your goodselves made it apparent that 'no difficulty baffles great zeal'.

There were some 21,000 gallons of milk at the premises of the 240 farmers who sent milk into our dairy on each day of this bitter period. Of all this quantity we are sure that not one pint of milk had to be thrown away. We realise that a great deal of credit must go to the farmer for the part he has played in making his milk available, but your assistance to these producers in your collection arrangements have not gone unappreciated, nor have the efforts sustained by your employees passed without admiration.

We would like you to pass our thanks on to all persons concerned with the handling of milk through this period.

Yours sincerely,

p.p. H. NICHOLSON,
Dairy Manager.

By all accounts, the winter of 1962-63 was as severe and long lasting as they come - so far as southern Hampshire was concerned. As the letter from the Cooperative Society indicates, the Hill milk collection drivers managed to get through all manner of snowdrifts, although Chris Hill recalls their Scammell Explorer recovery tractor was one of their busiest motors. It was used to pull many of their milk wagons through situations that would have tested the best all wheel drive vehicles never mind exotic cars like Jack Hill's American made Dodge. This winter was to be the swansong for Hills involvement in the milk collection trade for after more than 35 years of daily service, they liased with the Milk Marketing Board to discontinue their involvement.

A bad spate of accidents in the early 1960s (see opposite bottom) was one of the reasons why Jack Hill had decided to rationalise the company's activities. Hills involvement in cattle haulage ended on 18th February 1965 when the vehicles then in service were sold off as a going concern to EE Witchard of Alresford in north Hampshire. The transaction was made slightly easier as the cattle wagons had always operated under the licence issued originally to F Seaward Ltd, a company bought by Hills in the late '40s, albeit painted in Hills colours. Having regular drivers, the Hill cattle trucks were well known throughout the farming fraternity in southern England. As well as Fareham, Havant and Southampton markets, Hills also covered Winchester, Guildford, Salisbury and Chichester. Sheep to Devon and Cornwall was a regular traffic.

Hills bought two second hand Leyland Beavers - KOA 978 and 988 - which they operated originally on long distance general haulage hauling drawbar trailers. 988 was one of the 11 vehicles sold to EE Witchard as by '65, it had been transferred onto cattle work. Pictured leaving the Botley depot enroute for a load of pigs, Dave Hall usually drove this outfit. The Seddon four-wheeler seen behind the drawbar combination was one of six Hill vehicles with Neaverson meat boxes, which incorporated the latest type of meat handling, labour saving equipment.

BOTLEY HAMPSHIRE

(About 6 miles from Southampton, 12 from Winchester and 17 from Portsmouth,

immediately adjacent to Botley Railway Station)

JOHN JEFFERY & SON

are instructed by Messrs Hill & Sons (Botley and Denmead) Ltd.

to sell by Auction on the premises the

Extensive Range of Surplus

MOTORS, TANKERS and TRAILERS

on Thursday, 3rd August, 1967

commencing at 10.30 a.m.

On View Monday July 31st, 9 a.m. to 4 p.m.

Refreshments available — Licence applied for

Auction Offices, 9 Catherine Street, Salisbury (tel. 5337/8/9), Wilts.,
and High Street, Shaftesbury (tel. 2242/3), Dorset.

On Wednesday 2nd and Thursday 3rd August 1967, the Salisbury auctioneering concern of John Jeffrey & Son sold off more than 700 different lots of what was described as Hill & Sons surplus equipment, motors, tankers and trailers. The aerial view (overleaf) taken by Southern Newspapers shows the line up of vehicles in readiness for the sale and also the huge size of what is now the Bottings Industrial Estate. The first day saw 250 lots of miscellaneous equipment go under the hammer although it was the second day when the vehicles were sold which created the most interest. Best price was £1,000 which went to the 1957 Leyland Beaver UXC 957 (seen on the auction catalogue's cover) although it was probably the three compartment 3,120 gallon capacity stainless steel wine tanker which was hitched to it that created the value. Lot 286 was a 1957 Foden 2,000 gallon, 4-compartment tanker that went for £260. The Leyland Comet KGY 662 (which was sold for £95) had been fitted with a winch and converted for use by Jim Purkis as Hill's tackle wagon. It was new in 1949 as one of a batch of 14 similar Comets bought by National Benzole for operation as a bulk fuel tanker.

BRIDGE MOTORS

Up until the mid 1950s, the town of Botley hadn't been served by the most accessible of petrol filling stations so Jack Hill decided to fill that requirement through the construction of a purpose built service station on the side of the A334 road adjacent to Bottings Hotel. Naturally the building was constructed by Dudley Hill's concern of Hill Construction (during 1958) but as a banner for the new premises, the company of Bridge Motors (Commercial) Ltd had been created on 31st December 1955.

The Bridge Motors name was also used to cover the trading activities where all manner of plant, equipment and assorted vehicles were bought and sold. One of the earliest big deals involved the sale of 75 second-hand tankers to the Milk Marketing Board for the movement of skimmed milk (which was at that time considered to be a waste bi-product). Long after Jack Hill had handed down the reins of the heavy haulage work to his son Chris, he still enjoyed buying and selling things like second hand road sweepers to regular buyers in Italy, Malta, Africa and various other parts of the world.

The Bridge Motors name was also behind the construction of four articulated tractor units. These may have looked like Guys (and were in fact made up from Guy Warrior, Invincible and Big J parts) but they proudly bore the name of Hill.

The background to this quartet of Hill vehicles was changes in legislation. The mid 1960s saw the maximum weight limit for articulated outfits raised from 24 to 32 tons. However an odd quirk to the original regulations was the requirement for a huge outer axle spread for any four axled outfits.

In practical terms, this meant many operators were obliged to operate five axled artics if they wanted to take advantage of the extra payload potential.

Scammell were among the first to build their distinctive twin steer Trunker II tractor unit with this requirement in mind and Guy Motors at Wolverhampton also built a prototype 6x2 tractor unit. Jack Hill was keen to buy some of these Guy units but when their project was shelved, Guys supplied Hills with the drawings for the six wheelers and they were built at Bridge Motors (using second hand Guys).

The project almost backfired when the first Hill tractor unit - NOT 30F - went for its first annual MoT test. After operating at 32 tons gross prior to the test, the local officers of the Ministry of Transport refused to allow the haulier to operate this new 6x2 above 24 tons train weight. The reason they gave was that the name 'Hill' was not on the list of Approved vehicle builders. However, enquiries by Hills with the SMMT and the Motor Traders Association revealed that no such listing was required. It made matters worse when the local MoT personnel had the temerity to say that Hills had no right to build such vehicles. Saying something like was like a red rag to a bull so far as Jack was concerned.

An appeal was launched against the modest 24-ton rating. This was heard by a Ministry of Transport representative who arranged for the Hill / Guy tractor unit to be fully appraised and tested. This Inspector found in Hills favour in relation to the request for a 32 tons gross train weight plate. However, the MoT Inspector did ask for the original Gardner 6LX-150 engines to be up-rated so 6LXB and Leyland 680 engines were retro fitted. He also said the derogatory remarks about Hills ever building the vehicles were totally unfounded and he hoped the haulier wouldn't take the incident any further.

Bridge Motors was originally formed in 1955, as a trading name for the newly built filling station (see opposite) but it became the name behind all manner of dealing in plant, trailers and motor vehicles. The most impressive consignment arrived in August 1959, when 64 Hill trailers (of all shapes & sizes) were required to handle the shipment of car manufacturing machinery & civil engineering equipment from the USA. The earth scrapers had been shipped in two parts but were re-joined prior to being towed to the Botley depot for re-painting. The equipment & machinery were delivered to Harper Engineering's premises at Exeter Airport while the scrapers were sold & delivered to customers all over the country.

In the late 1960s, you could say that Hills had a love affair with the Guy product and made use of it in a huge variety of forms. Apart from the badge, there seems little to differentiate a Hill tractor unit from a Guy Warrior / Invincible / Big J eight wheeler, without its second drive axle. The Hill units were built in the company workshops and those involved in the work included Jim Purkis, Joe Appleton, Phil Bryan and Charlie Knight. The units were made from second hand purchases, Hill no. 3 being originally the Guy tanker 835 DBE. Given new registration, NOT 30F was the first of these special six wheelers. It's seen with driver West-Thomas fixing the side markers. And perhaps only Hills would ever think of using a Guy Invincible cab as a shelter for their girder trailer steersmen.

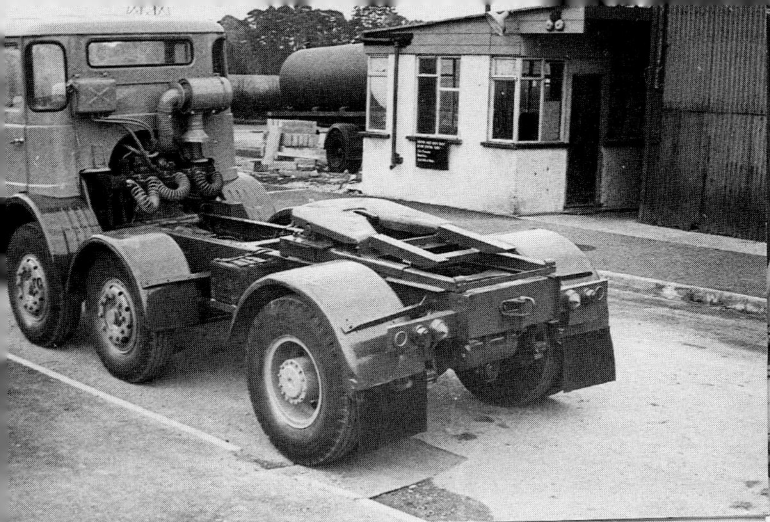

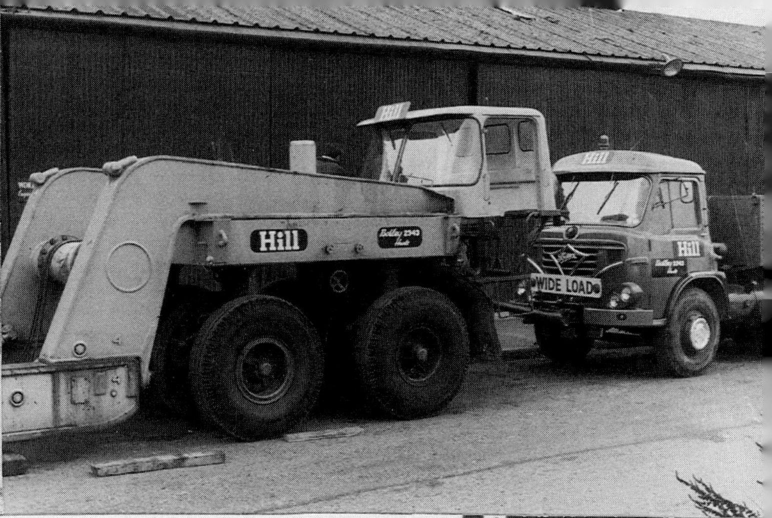
Hill
WIDE LOAD

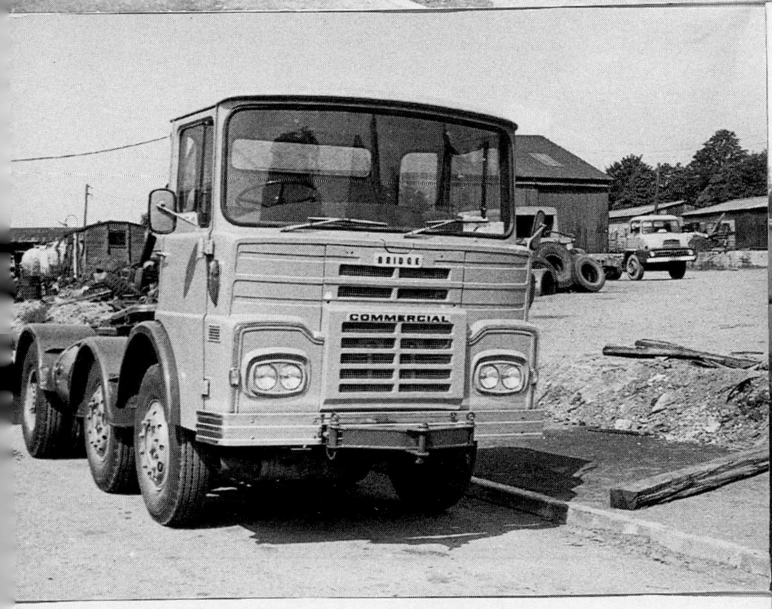
BRIDGE
COMMERCIAL

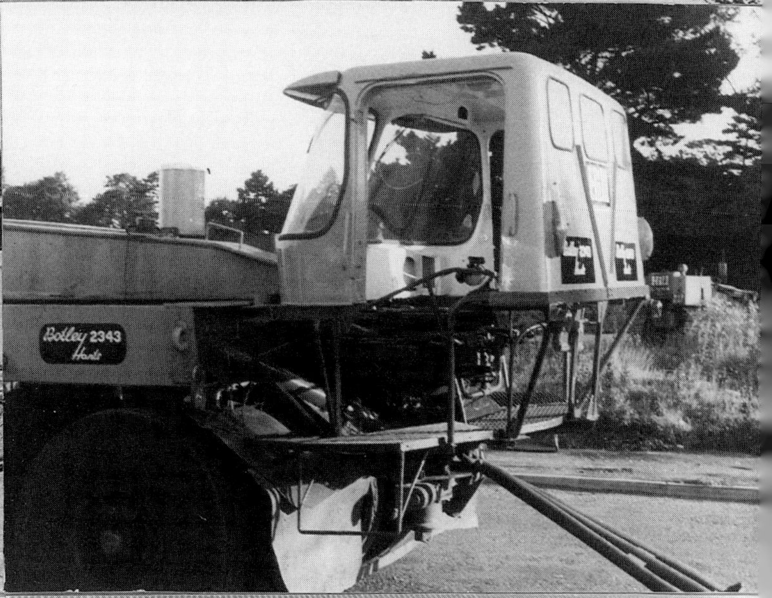
Bolley 2343

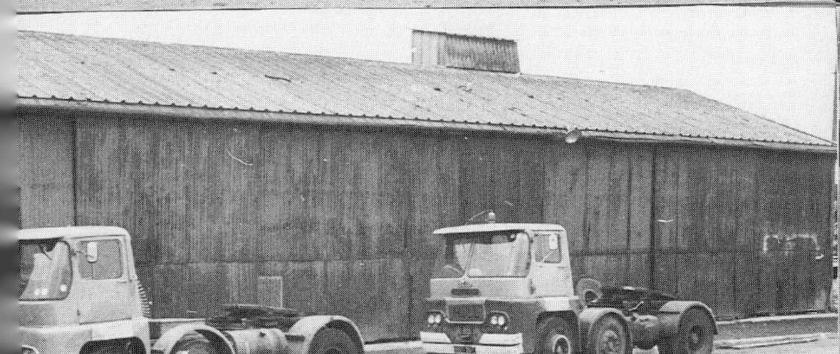

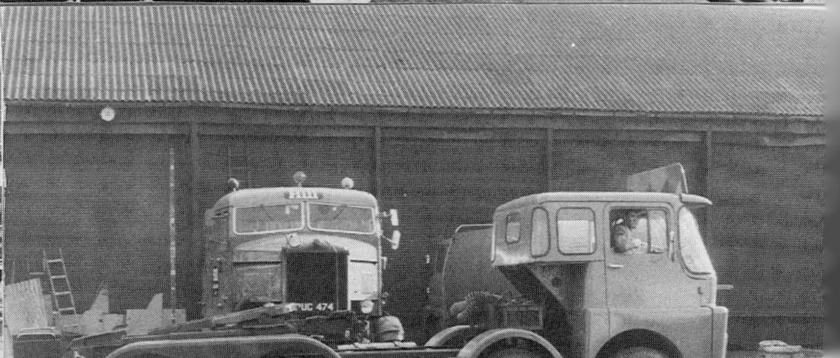

Hill
Hill
Bolley 2343

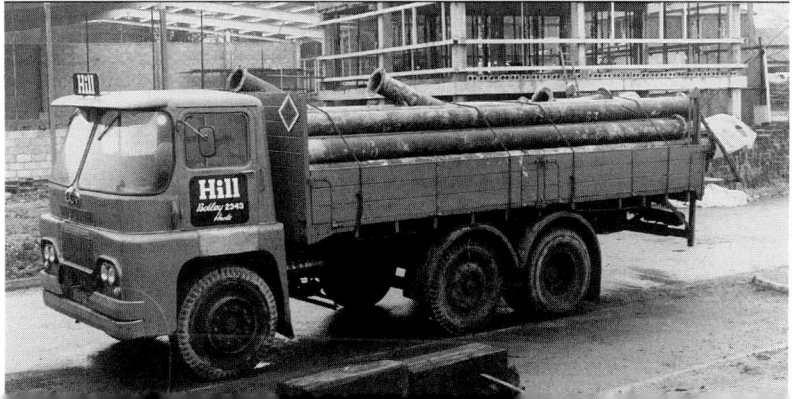

Jack Hill found this strange looking tank transporter at a quarry in Gloucestershire. While the 4x4 tractor unit is something of a puzzle (and Jack didn't buy it) the 30 ton Scammell trailer was pressed into use when Hills had a hurry up job demanding they move 60 Sexton tanks out of the Bicester tank park within 10 days. A conventional Scammell Pioneer was used to haul this trailer for that work.

These Portsmouth trolleybuses were two of the six which were withdrawn by the local Corporation, stored then sold in April 1960 to the George Cohen 600 Group. They were scrapped by Pollock & Brown at the Northam Iron Works, Southampton in July 1960 and Hills were contracted to haul them on their last trip.

Direct Roadways Ltd consisted of a fleet of four vehicles, which Jack Hill bought mainly for their 'A' licences in 1960. This Albion Clydesdale was subsequently repainted but it was sold on - again with its licence - in May 1966 to Rosemary Searl for £2,750.

With the Botley yard stretching to 14 acres, it was a haven for all sorts of vehicles and loads. The Federal (top) was bought from the Ruddington sales and while Jack did have plans to use it, the vehicle was eventually sold on. Phil Moth was able to furnish background to the Sadler Rail Coach. In 1964, Charles Ashby leased the part of the recently closed Meon Valley railway line in Hampshire running between Knowle and Droxford. His intention was to conduct trials with his 'Pace-Railer', a luxury four-wheel rail coach that had 50 reclining seats and air conditioning. He had hoped to use the vehicle - and others - on scenic rail routes but when the coach was severely vandalised, the idea was shelved.

Chris Hill was a regular visitor to the Botley yard and he's seen polishing his drawbar trailer shunting technique with the yard tractor while Joe Appleton looks on. A fifth wheel dolly converter was also used so the tractor could shunt any semi-trailers. Simmons Engineering of West End, near Botley, converted this DG Foden heavy hauler (centre) to a yard crane. One of the many strange vehicles to come into the Hill yard was this unregistered left hand drive 'Tin Front' AEC Mammoth Major eight wheeler (bottom). Jack Hill actually considered putting it to work but it was eventually sold abroad.

While modern day craneage is almost taken for granted, more than 40 years ago, Hills had to use more basic methods of lifting. The ex military AEC 0854 six wheeler (with the cab & front end of a Matador, plus rear bogie from a Marshal) which Joe Appleton often drove, proved a willing workhorse although as yard crane it was often totally abused. The crane was nominally rated as having a 6 tons capacity but had its life extended in 1973 when the front half of a Guy eight wheeler was transplanted into the AEC's rear half. ROT 428G (another of the Hill / Guy specials) is having its load repositioned after striking an overhanging tree close to the Botley depot.

LONG, WIDE AND AWKWARD

It was always going to be a tight squeeze but even with the long arm of the law offering assistance, this right turn into the High Street at Fareham proved impossible - until one modest road sign was removed. Driver John Murphy and mate 'Macca' McAlpine were taking the first of eight similar 100' crane gantry sections from Littlehampton to Dunfermline on Thursday June 29th, 1961. Fareham photographer AE Ayre was on hand to record the hold up, which stayed in place for about an hour. Some smaller cars were able to squeeze under the back of the load but the problem was only solved when a bulldozer pushed the offending sign out the way. By then, every motorist in the area knew the name Hills of Botley.

During the construction of the Alloa power station at Long Gannet in Scotland, Hills undertook the transport of 17 loads of boiler expansion joints from Vokes steel factory at Guildford Park. Some of the fabrications were 17'5" square and because they had to pass under a 17'4" high bridge at Alloa, Vokes fitted out this Hill trailer so it could carry a large number of frames at the same time - and still beat the height restriction. While the trailer was moved internally with a box tractor and a dolly axle converter, driver George Holbrook and his Scammell Highwayman artic did most of the runs to Scotland. The work was done over four years and for each load, apparently 68 notifications had to be sent out to the relevant interested authorities.

HILL
BOTLEY
HANTS
269

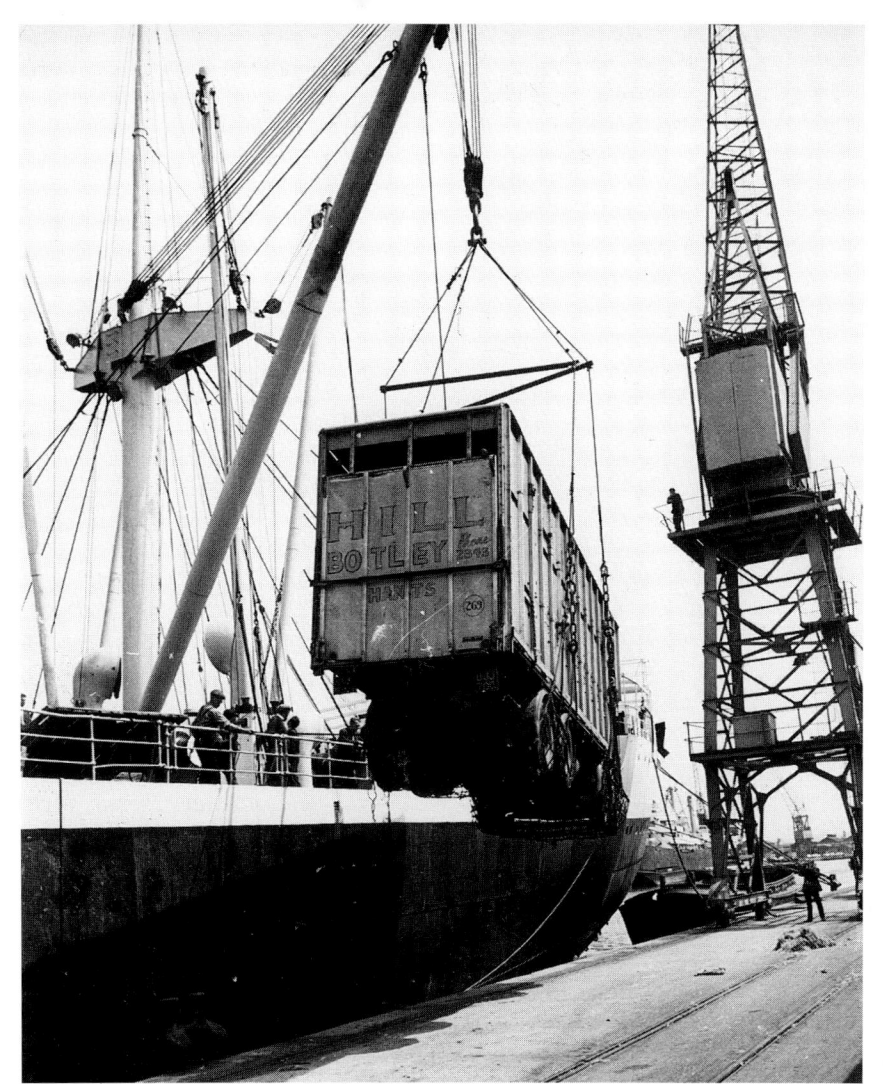

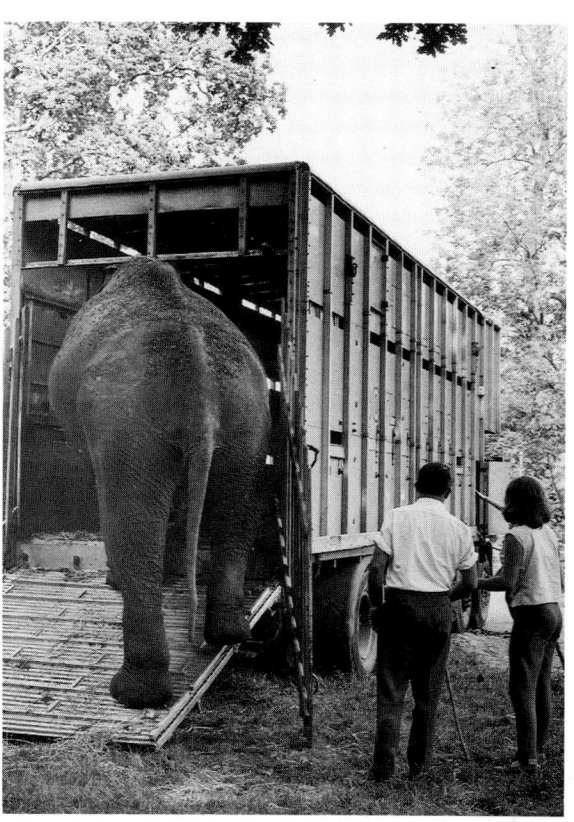

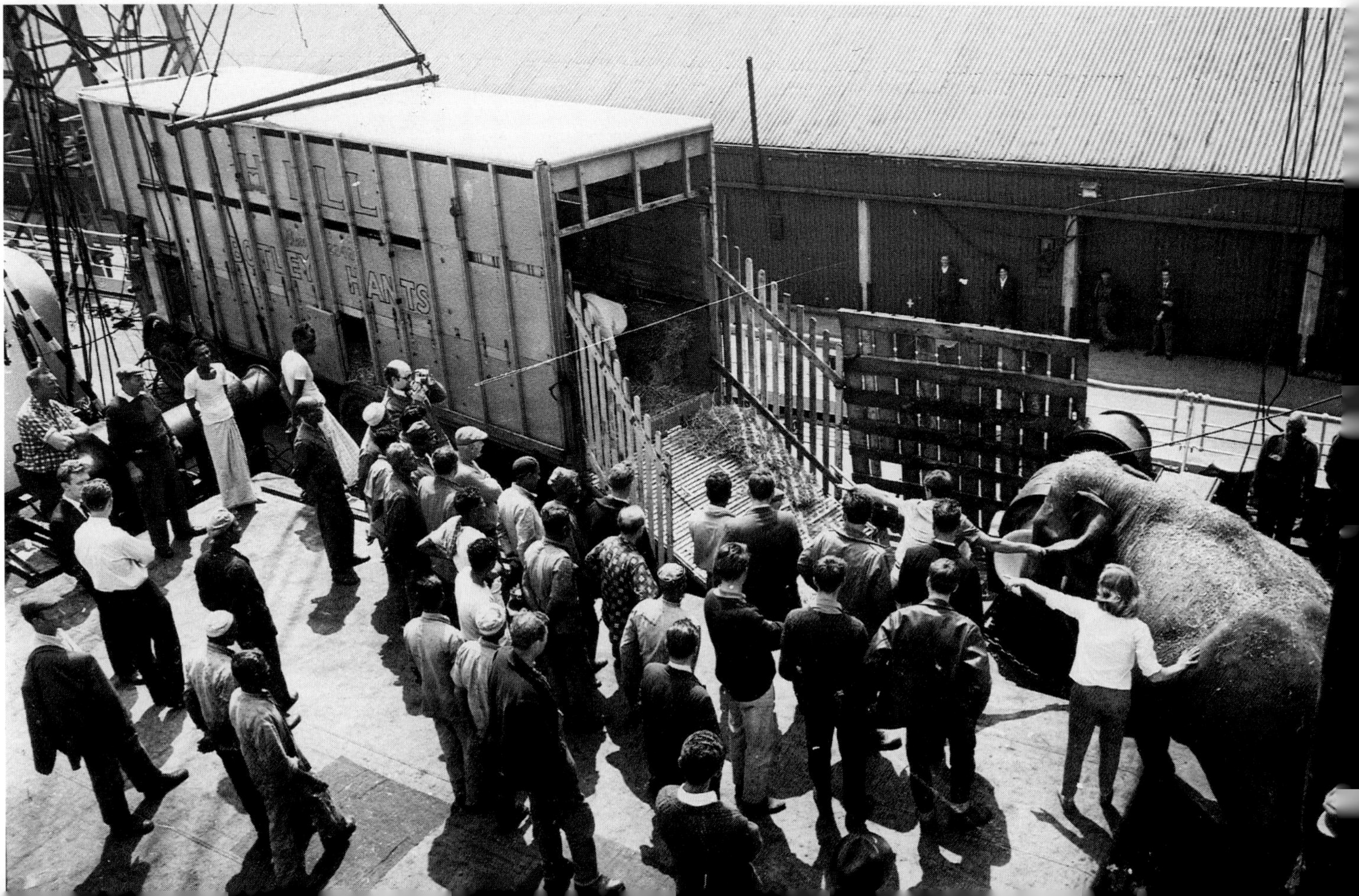
HILL
BOTLEY
HANTS

Dealing with animals was something of a Hill speciality although one of the most awkward was the 14-year-old Indian Elephant named Zubanda. It was brought to London by ship but refused point blank to be lifted onto the quayside by slings. The alternative was to lift the eight-wheeled Leyland Octopus onto the boat and then coax the elephant on board. Long serving Hill driver Len Samways made the trip to Southampton Common quickly enough but because Zubanda was so large, it couldn't turn round inside the eight wheeler. Mary Chipperfield had to coax the elephant backwards and there was a huge sigh of relief when it reached terra firma. No such problems with the pictured giraffes (leaving the MV Chindwara in June 1959) while Jack Hill recalls that when they moved some monkeys, these animals even helped the Hill crew to secure the load by stretching their arms through the cage bars to pull the covering sheets down.

If Hills had a speciality, it was in being able to make the best out of all manner of military surplus equipment - in all manner of forms. The Leyland Comet LOR 712 (opposite top) is enroute to the 1964 January boat show at Earls Court with a rather special Proctor boat mast (destined for use in the America's Cup race) but taking the weight at the other end of the extendable pole trailer is the running gear from an ex military tank transporter. Hills had a workshop staff of 17 (plus four carpenters making bodies etc) and it was the welding talents of guys like Jim Purkiss that was able to purpose build trailers for the most awkward of loads like the one (bottom) at Always Welding, Chichester. The Thornycroft Antar (opposite bottom) is pictured leaving IPV at Eastleigh with an 80' pressure vessel heading for export to Istanbul.

In late February 1965, Charlie Tigwell and XNM 477 (accompanied by mate John Farringdon) moved this 100' crane gantry beam (centre and top) from Fratton rail terminal at Portsmouth to a new factory being built for the Halmatic boat builders at Havant. Charlie is the figure on the left (bottom) at Cardiff with an earlier Scammell and mate Bill Miles, the figure on the right.

Described as being particularly ticklish was transporting the main plane of the TSR 2 from Boscombe Down to Henlow during 1968. George Holbrook was driving with mates Barry Hill, Chris Hill and Joe Appleton. The use of an adjustable tilt frame allowed the Hill crew to raise - or lower - the height of the load when they encountered immovable objects like buildings or bridges.

Both Hills and artic driver Freddie King were fined (£25 and £5 respectively) for insecure load after this incident in Wickham Square during August 1968. However, when the case went on appeal to Hampshire Quarter Sessions, the convictions were thrown out because of the special circumstances. It was accepted that Hills had safely moved 3,400 of these 35' long, 5-ton natural gas pipes while they were supported and restrained in the same manner. This incident only occurred when driver King had to swerve to miss a car coming out from a minor road. The car driver was fined for careless driving and because the local police confirmed there had been no problems with any other load of pipes, the case was dismissed. As a footnote to this incident, Freddie King said that as soon as the Albion came to a halt, it was the quickest he'd ever got out of the cab.

Freddie King and his Albion Clydesdale (which was new to Hills in 1964) are seen in happier days involved with pipes large and small. The large vessel (bottom) is seen in the premises of Scotts at Littlehampton during March 1966. The smaller drainpipes (top) were taken by Freddie and the Albion from the West Drayton, Middlesex works to sites all over the UK.

HEAVYWEIGHT SPECIALISTS

During the 1960s, Hills used all manner of other people's hand me downs as their flagship heavy haulers. The Scammell Mountaineer 298 FUW started life with the Royal Navy as an artic. Hills used it like this for a short time but found its heavy unladen weight in artic form (13 tons 5 cwt) meant its road tax was excessive so they converted it to a ballast tractor. Together with regular driver Titch Nutley, it moved a huge number of concrete beams and is pictured with the last Anglia made beam for the Windsor relief road (opposite bottom). Jack Hill's brother in law Bob Darsley is the figure in front of the Scammell. Hills preferred choice of bogies were cut down ex military tank transporters (known as SMTs) also modified with the fitment of manual steering gear.

298 FUW

HILL

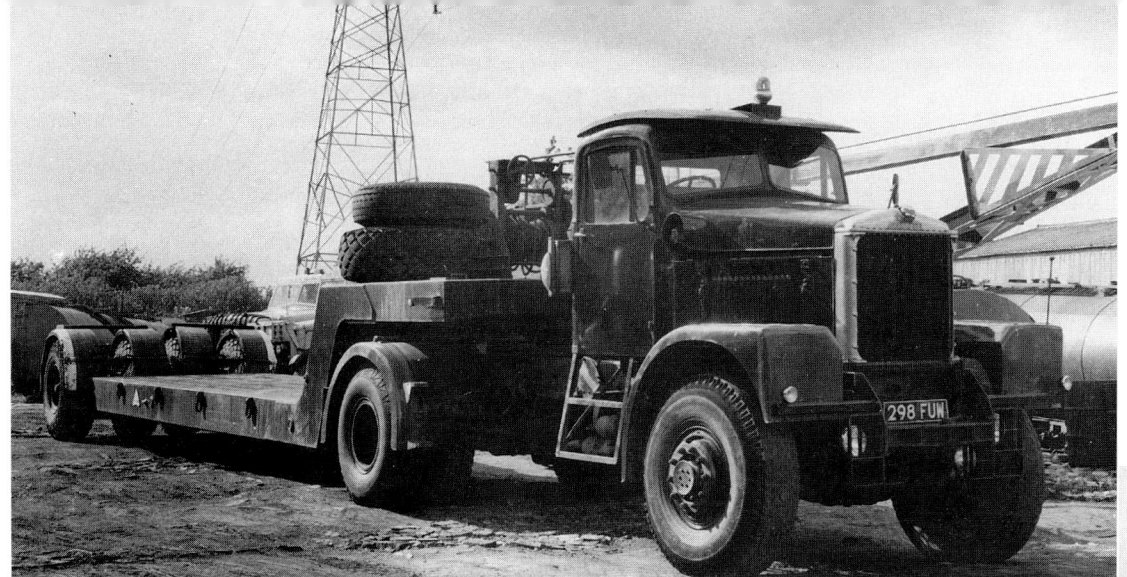

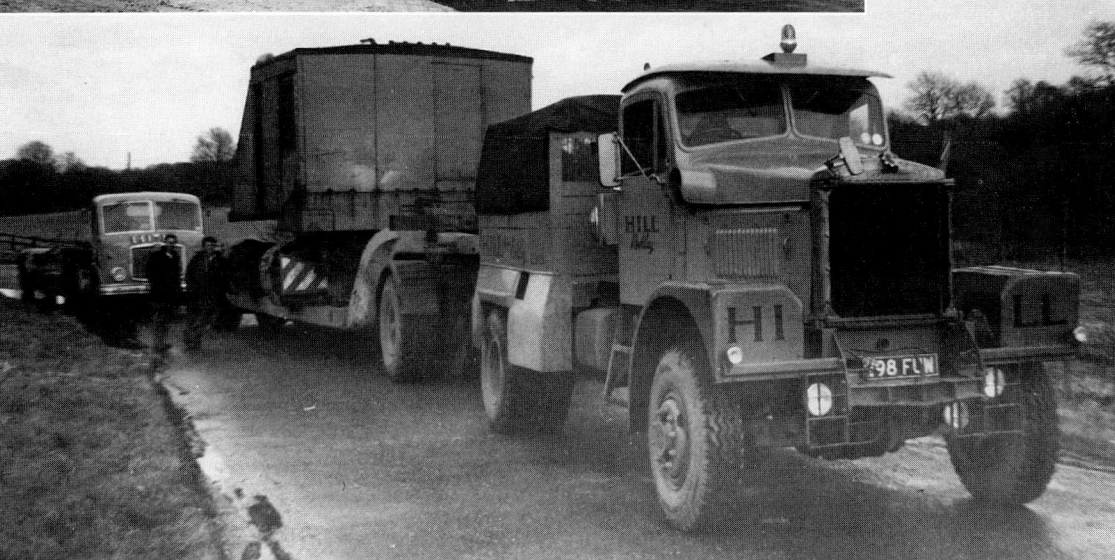

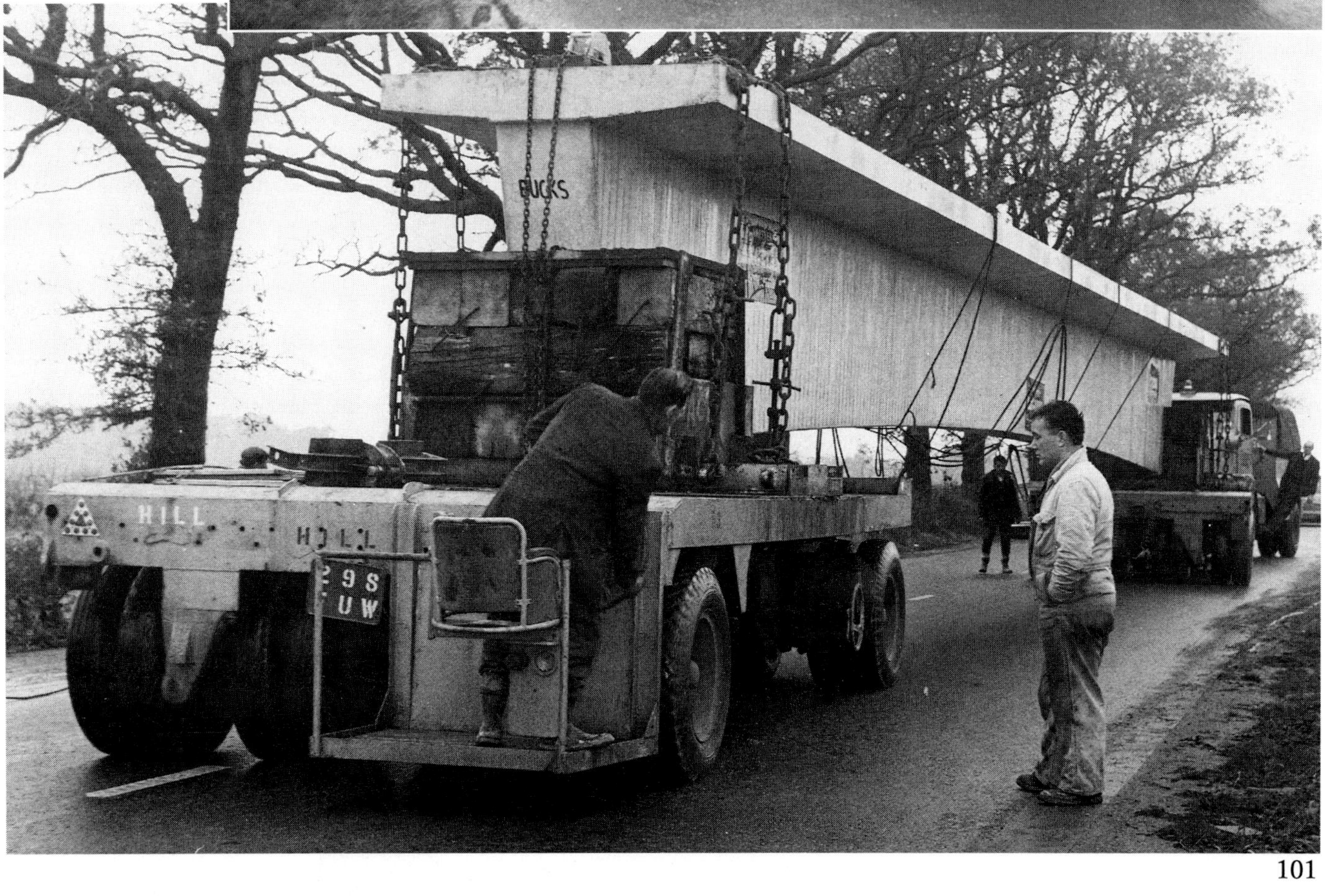
BUCKS
HILL
HILL

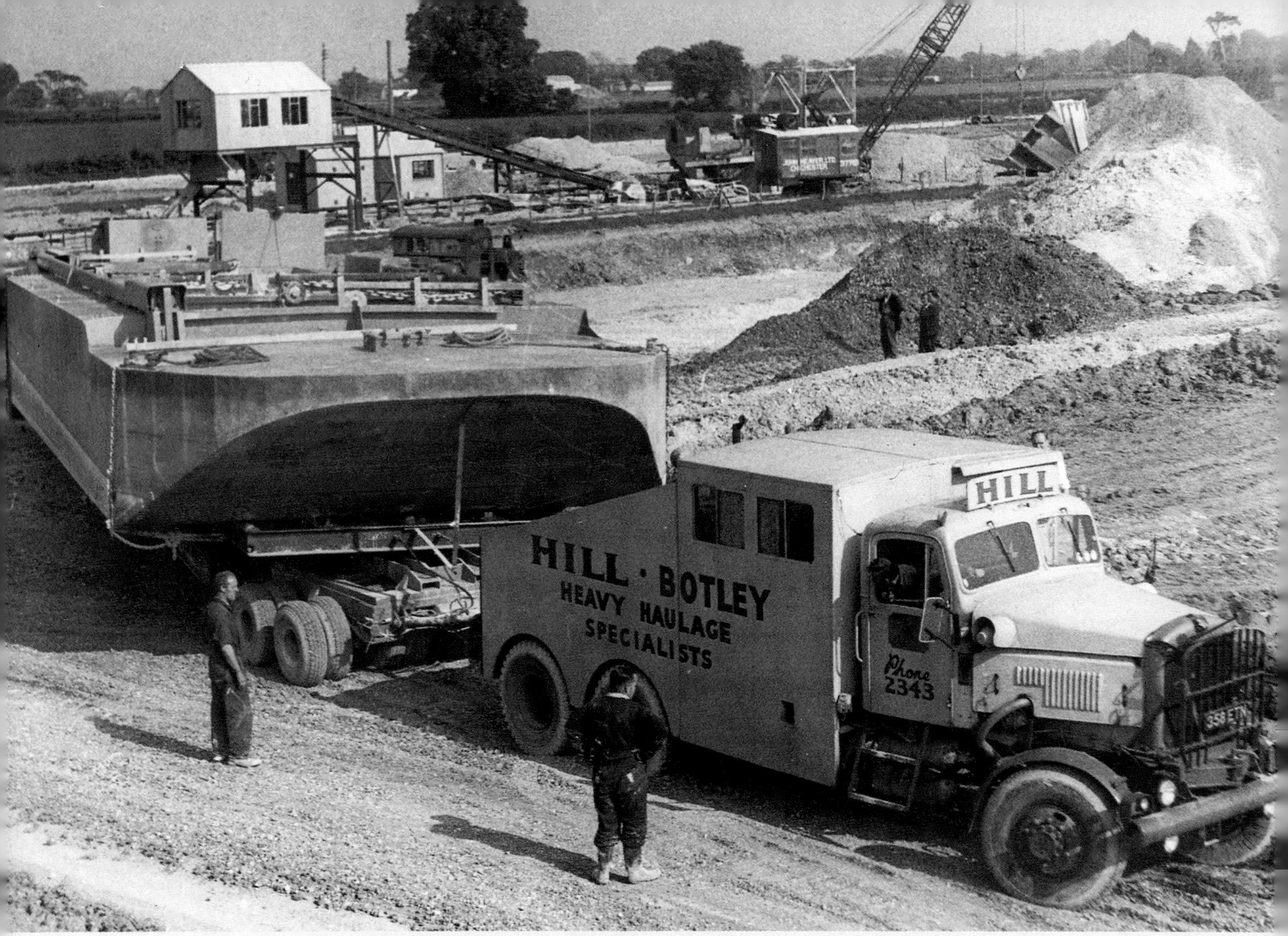
HILL
HILL · BOTLEY
HEAVY HAULAGE
SPECIALISTS
Phone
2343

HILL
HILL
Botley Hants

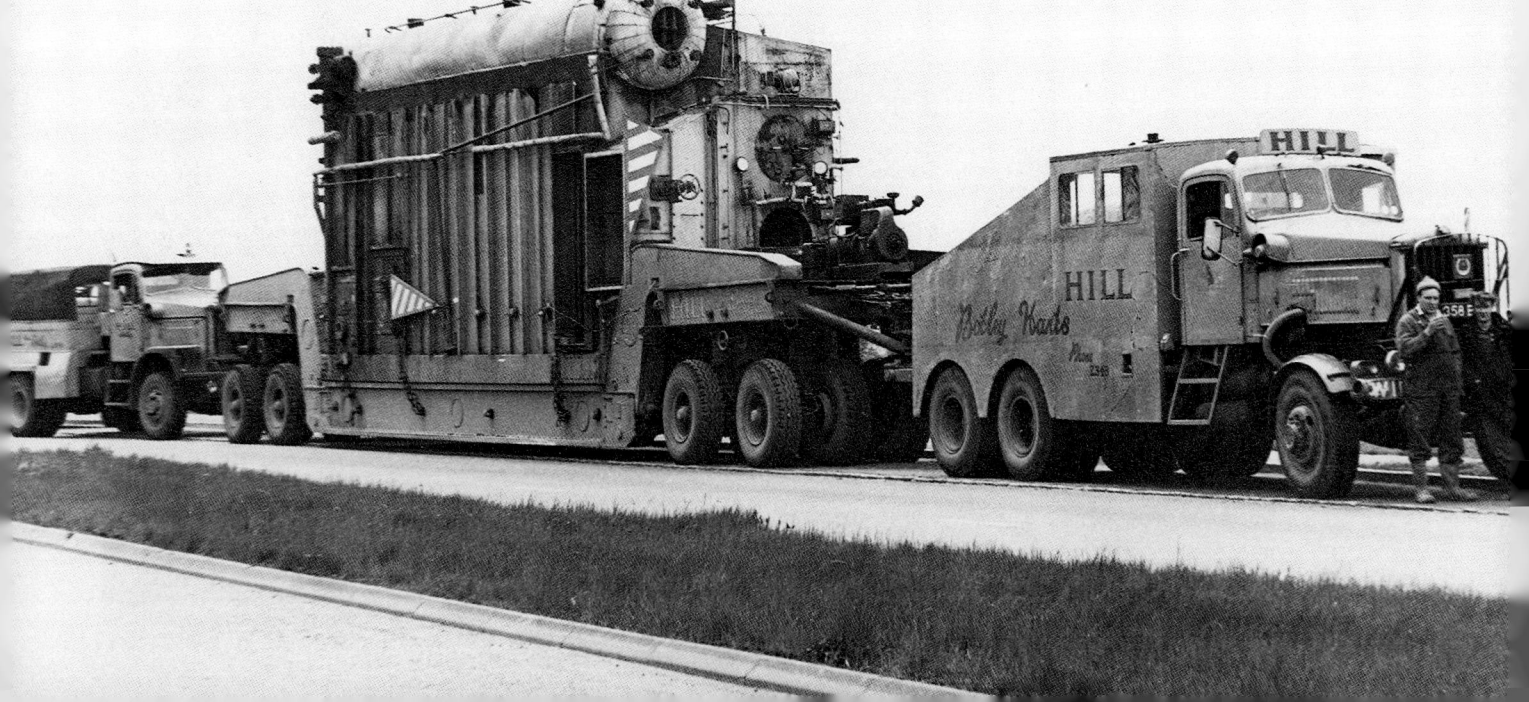

The 6x6 Scammell Constructor 358 ETN started life in Northumberland as a Reich drilling rig carrier. Bought by Hills, it was given a fresh coat of paint (plus an extended crew cab) and expected to do all sorts of jobs. In May 1963, it made easy work of moving this barge (opposite top) across the road from the quarry at Portfield where it had worked for the previous two years (see also pages 71, 72, 73). The haulier had to make extra long

girder beams to haul this 18' high Foster Wheeler package boiler from Stanford-le-Hope in Essex to Stockton during August 1967 (opposite bottom). Gordon Prebble and Barry Johnson are amongst the crew. The 54RB excavator (bottom) is seen in the AJ Lennie photograph enroute to Trawsfynydd in North Wales. Part of the route took the load over a bridge so weak, the trailer had to be winched over separately of the two tractors involved. Jack Collins was believed to be driving TOT 297, the pusher at the rear.

XMT 586 was a specially built narrow version (9' wide) of the 6x4 Thornycroft Antar which was bought new by the civil engineering concern Sir Robert McAlpine in 1959. Correspondence with Hills dated November 1960 reveal that the haulier was offered this vehicle for £3,500. It worked hard for Hills (until sold to a showman about 1970) although it did have problems with its Rover Meteorite diesel engine. Jim Purkis recalls the Antar as being: 'A blooming good motor with fantastic brakes.' It was usually operated with a girder trailer carrying excavators all over the country. When heading for Wales or Scotland, an extra pushing tractor was used. Titch Nutley and Jack Murphy also drove this tractor while the mate on the running board (centre) is Charlie Samways.

The 1960 Scammell Junior Constructor 874 AUU was chassis no. 10689. It was bought new by the contracting concern Marples Ridgway & Partners specifically to move the 14,000 tons of concrete (divided into 906 units) involved in the construction of the Hammersmith flyover in London. Once the work was completed, Jack Hill bought the Scammell - and two Crane drawbar trailers - in November 1961 for £4,500 plus £6,500 for the trailers. It proved a versatile tractor for Hills and was used on a variety of work. For the Gleeson machine it required the assistance of a Scammell Pioneer (on Winchester by pass) and especially when climbing the fearsome

stretch of A6 over Shap (insert). Hills converted this ballast box tractor for artic use and it's seen in that form at Basingstoke with driver Peter 'Birdseye' Howlett waiting for a police escort. Hills worked it until about 1975 when it was sold for export to Africa.

There wasn't a lot of weight in the three loads of 82' long steel girders which had travelled by rail from Middlesbrough to Fareham station in early January 1970. But, what made the job interesting for Hills was squeezing into the entrance of the Vosper Thornycroft's Woolston yard at Southampton - without striking the buildings opposite. The girders were for a new specialised facility for building naval vessels in glass-reinforced plastics. John Bird was on hand to photograph the moment while the Hill team included Maurice Adams, George Holbrook and Monty Theobold as one of the rear bogie steersmen. Hills bought this ex Pickfords 6x6 Constructor during April 1968 (for £1,600) and while it is registered PUC 474, it was actually rebuilt using some parts from the similar ex Pickfords Constructor PUC 472. Perhaps the most impressive job involving this 6x6 was when it worked together with the 1955 ex Siddle C Cook Constructor SPT 600 (which was bought by Jack Hill on 17th May 1968 for £925) to move two impressive fractionating columns (see opposite and following two pages).

Birkenhead L41 9BP England

Telephone 051-647 7080
Telex 62463

Hill & Sons (Botley & Denmead) Ltd.,
Bottings Hotel,
Curdidge,
Botley,
Hants.

Our ref	MAN/WOR/AH/LHN
Your ref	
Date	8th July, 1970

For the Attention of Mr. J. A. Hill

Dear Sirs,

MOVEMENT OF 2 LARGE PRESSURE VESSELS FROM CANUTE ROAD RAILWAY STATION TO ESSO REFINERY, FAWLEY

We are writing in connection wth the loading and transport of the above 2 very large Pressure Vessels from Canute Road Railway Station, Southampton to Esso Refinery Site, Fawley on Saturday and Sunday, 27th and 28th June, 1970, and would like to place on record our thanks for all the help and assistance you and your staff gave us to complete this rather difficult and hazardous movement so successfully,

We would mention that the Ministry of Transport and the Railway Company had been informed by a well known haulage contractor in Southampton that the larger of the two vessels could not possibly be moved out of Canute Road, Southampton by a road Vehicle.

Thanking you once again for your very efficient assistance in loading and transporting these very large vessels from Canute Road Station to Fawley Refinery Site.

Yours faithfully,

107

The Hills crew show how to do it - and also show how not to do it. The Crane girder trailer (top) supports a press, which was removed from a factory in Southampton and taken - via the Red Funnel ferry - to the Isle of Wight. The trip across the water was bad enough with concern over the load's high centre of gravity but when the Hill crew arrived at British Hovercraft's premises in Newport, the severe incline of the access road prompted the strange position of the trailer. Packing the trailer up in this fashion meant one bogie was clear of the ground but it allowed the press to be winched off horizontally. Amongst the Hill crew were Barrie Johnson, Maurice Adams and Titch Nutley. Note should be made of the city names: Rotterdam, London & Paris painted on the trailer necks. You'll have to ask Chris Hill why that was done - although he may not tell you. The Scammell Trunker came to grief in Southampton docks because the crawler crane was slewed when not secured to the tandem axle King low loader. No damage was caused although another crane was needed to restore the balance of the stricken Pennine.

RUP 900 was naturally nicknamed 'Rupert' when it came to Hills although it had been named 'Smokey' when it was with its first owners, Siddle C Cook of Consett. Driver John Panton is pictured in Hill's yard checking out his mobile crusher load. The 1955 4x4 Scammell Mountaineer was written off in the early 1970s after an accident while working on a pipeline contract.

The 240-ton Scammell Contractor TRL 924H (ch.no. 4024) was built in 1968 but not registered until January 1970 when it went to the Mid Cornwall Contractors division of English China Clay. Originally operated - and sold - in artic form (bottom) it had only covered about 50,000 miles when Jack Hill bought it in February 1972. Complete with King 65 ton capacity low loading trailer, the asking price was £10,500 although £5,100 of that was offset by the part exchange of a brand new Guy Big J tractor unit. Pictured at Slough railway goods yard, the Brighton Belle coach it's about to move was destined for the Thorpe Restaurant in Farnham. Hills sold this 6x4 Scammell in 1976 to Magnaload although the extra crew accommodation (believed from a D8 Caterpillar bulldozer) was removed before the vehicle made the long trip north to Teesside.

A number of Scammell Crusaders joined the Hill fleet in the 1970s and these included two Detroit powered 6x4s which were rated for work up to 65 tons gross on heavy haulage work. Jack Hill paid £4,320 for the left hand drive 1970 model SAR 457J when bought from Roebuck Truck Sales in Wolverhampton in July 1975. SNB 431J cost £7,500 in June 1972 when purchased from Brinksway Motors of Stockport. This tractor was new to Avis Truck Hire in 1971. The Crusader is seen coupled to a King low loader at Botley (top) with Chris Hill, the figure on the left, talking to driver Bob Barnard.

Jack Hill bought all sorts of vehicles & trailers either for his own company's use or to sell on. However, one concern which didn't wish to deal with him was Pickfords and when the 240 ton Contractor SYO 376F came to Botley (in January 1974) it was bought via Wesley Turkington of Lurgan in Northern Ireland. This 6x4 tractor was new in October 1967 and Hills paid £9,102 for it. It's pictured moving a Crane girder trailer, which was bought from Hardwicks of Ewell (but started life with Sunter Brothers) and is being hauled to the docks for export.

Hills made use of an assortment of Foden built tractors and those of the marque in service during the 1970s included the ex Everley Bros of Hayes PYH 546F (centre right) that was worked in both ballasted and artic tractor form. Seen coupled to a King 60 ton low loader, driver Vic Bridges and mate Len Page are securing the Frankipile Piling machine. The S40 cab'd UFU 937J (centre left) was new to St Vincent Plant of Scunthorpe in August 1970. Chassis number 70058 had a Cummins 250 engine and was rated as a 75 tonner.

One Foden tractor which you won't find in the Hills fleet list is this one. Bought brand new through Scotts of Nottingham in August 1974 (for £17,135) it had Cummins 335 engine, torque converter and a plate describing it as a 100 tonner. It also came with such a poor steering lock that Hills decided to sell it on before it was used or even registered.

The war time midget submarine X24 (unofficially named 'Expeditious') had laid in the Royal Navy's submarine headquarters at Gosport - HMS Dolphin - from 1948 until March 1973 when it was decided it would make a fine exhibit in Portsmouth dockyard close to HMS Victory. As the crow flew, the sub only had to travel 1,000 yards. But because its disrepair meant it was in no condition to travel there under its own power (by sea), low bridges in Fareham forced it on a 47 mile round trip through Southern Hampshire. Hills used one of their two ex Cardiff Plant Hire Guy Big J 6x4 tractor units for the job which George Holbrook drove. The CS Dalby photographs show the semi-trailer's rear steersman was Jim Purkis who also had a hand in building this particular 70-ton capacity trailer. It had started life with Sparrow Crane Hire of Bath - as a drawbar trailer of 140-ton capacity - but Jim's expertise saw it converted into two 70-ton semi-trailers. Once at Portsmouth dockyard, it was planned that X24 would be fully refurbished but would also have part of its hull cut away so that visitors could see inside it.

Built by Fairey Allday Marine for Hovertravel, the AP 1-88 was bannered as a new generation of all welded, diesel powered hovercraft when it was launched on December 21st, 1981. The craft was partially built in premises at Fort Brockenhurst but was fitted out after being launched then towed to the company's premises in Bembridge, Isle of Wight. The Hill crew involved in the short but delicate road move were Jim Purkis, Len Page and Keith Knight. They had to contend with a film crew from TV South getting in the way plus two occasions when champagne was splashed over their load. Hardest part of the job was probably just getting out of the construction shop as the 71' long, 33' wide vessel had to come through doors which would only open to 34' wide. The Saturday was spent getting the 10.5 tons craft outside and once extricated, Peta Sutherland, wife of the MD of Fairey Allday, broke the first bottle of champers after a short 'Bon Voyage' speech. The next day, Sunday, saw the outfit travel at walking speed along the A32 to the Royal Naval Aircraft Yard, Fleetlands, where it was lifted into the water (the following day). Once launched, Chris Bland of Hovertravel splashed another bottle of bubbly over the hull. The AP1-88 had four Deutz engines - two 8 cylinder 324bhp ones to provide lift and two 12 cylinder 428bhp ones to provide propulsion. Capable of carrying 80 passengers, it was scheduled to operate on the Ryde - Southsea service.

This 106' long wooden footbridge was hauled from Devon to Woodley at Reading during July 1984. At the time, the Scammell S24 6x4 tractor HBM 529Y was an example of the latest and greatest from the Watford manufacturer but to support the rear of this £25,000 structure, Hills preferred their own special bogie. Jim Purkis had worked his magic on a vehicle which started life as a Guy Big J 6x4 tractor unit (to Cardiff Plant Hire) and it proved so versatile, it was still in the Hill trailer park in 2001. Nobby Brown is steering the six wheeler without the engine.

Hills were to operate a pair of Scammell S24 6x4 tractors of 150 tons design capacity. LTR 689Y didn't stay too long at Botley before Tony Marler of Staylybridge, Cheshire offered Hills a price they couldn't refuse. The S24 did return to the Hill yard when Marlers put it up for sale (some time later) and while Hills considered it a good motor, it was eventually exported to Malta.

The old Guy Big J six wheeler was called into action to support the rear of these 80' trusses (one of four loads) which were to form the roof of the newly constructed Chichester swimming pool. Made in Sweden, the MAN collected them from Shoreham docks. Keith Knight is the driver while Big John Evans is the mate. The late '70s and early 1980s saw Hills take eight MANs into service (because of their general reliability). Chris Hill recalls using the bonus loader type of jeep dolly to quickly convert his 4x2 units into 6x2s - until certain enforcement agencies decided to be picky about the fine print of the law.

9th August 1985 was quite a day for Richard Branson and the Cougar Marine concern when the 'Spirit of Britain' was launched by Prince Michael of Kent at Port Hamble, Southampton (opposite). There was a band and a massive crowd on hand to wave to the invited dignitaries while Hill's S24 Scammell and driver Keith Knight took centre stage. The Scammell had drawn the vessel out of the manufacturing shop into the daylight on its special launch trailer and was also used as a land anchor during the launching ceremony. Because the event had been well rehearsed, it went off fairly smoothly. The vessel was the Trans-Atlantic challenger for that year although it subsequently didn't fair too well in the race. Hills involvement with this catamaran had began a year earlier when they'd moved the hull from the Ford aerodrome (where it had been built) to the Avon river at Littlehampton where it was launched into the water. Dick Lodge is at the wheel of the Scammell S26 (above) running in ballast tractor form with the craft being supported on two of Hill's special bogies. Due to its excessive width, the hull had to be lifted over an entrance gate and reloaded onto the Hills outfit before travelling another 100 yards to the quay. After being lifted into the water, it was towed by sea to Port Hamble prior to Cougar Marine getting to work on it.

Hills long standing expertise in moving battle tanks and similar military crawler driven machines is still regularly called upon. A conventional Trailmaster low loader was sufficient to handle the two Scorpions behind Alistair Brown's MAN enroute from Aldershot to Tidworth while the Scania 143 (bottom) is using a highly versatile Nooteboom semi-trailer. Called a 'Multicarrier', the all wheel steer five-axle trailer can handle payloads up to 90 tonnes, which is more than enough to carry the latest Challenger II tanks. It also has a double extendable chassis, which will increase the trailer's length to about 86'. The two stage hydraulically operated loading ramps are fully detachable. It was a break in military movements (while the British Forces were all overseas) that allowed Chris Hill to have Miles & Nunn of Whitchurch in Hampshire add a fourth axle to the company's 143-450. It didn't change the Scania's 150 tonnes gross rating but it gave greater peace of mind with axle loadings.

At 60 tonnes in weight and up to 36' high, Hills sub contractor Ivor Truman recalls moving these parts of dockside cranes on Canary Wharf in London during April 1995 as a testing job. Ivor had changed his Foden for this ex Interlift 150-ton rated Daf 95 6x4, which was being mated by his son Richard. The Truman team moved these cranes twice. They were taken about quarter of a mile on the King four-axle low loader to be shot blasted and re-painted before being returned to where they'd started. Ivor had spent 32 years on the road but in 1996 he decided to hang up his keys. He's still involved in dealing with abnormal loads as part of his work in Southampton docks.

Driver Stuart James and mate Roy Wiseman took this cold box (top left) from Wrexham to Southampton in March 2001. Overall length was 102' while gross weight was 94 tonnes. There wasn't a lot of weight in the building (above left) which Keith Knight, Roy Wiseman and Chris Dugmore shifted during December '00 at Northam, Southampton. But at 83' long and 26' wide it required the trailer's hydraulics to first lift it. It was pulled forward but then lowered onto blocks so the Iveco could pick it up from the other end. The building was then taken three quarters of a mile where it was skated off onto plinths that were built to receive it. Moving the building in one piece saved the contractors the problem of taking it all to bits (and then having to re-build it).

You have to look at the Bottings Industrial Estate from the air and then compare it with the other fantastic aerial shots on pages 32, 80 and 81, to fully appreciate the changes that have been made over the last 70 years. As the Hills operation has retracted from its peak in the early 1960s, the land around their depot has been sold off and developed for other uses. In 2001, the sign above the big garage remained as a last indicator of the Estate's proud founders. While who but Hills would ever think of utilising the space on a roundabout (opposite bottom) to use as a trailer park

*Building
and Constructional Engineers*

Telephone
Southampton 54022-3 (2 lines)

Works Address :
COMMERCIAL ROAD
TOTTON, Hants

**CONSTRUCTION COMPANY
(ENGINEERS) LTD.**

Head Office:
WIDE LANE
SWAYTHLING
SOUTHAMPTON

Hills of Botley weren't the only Hill concern to make their mark in Hampshire. By the time Dudley Valentine Hill sold out to the Atcost concern in May 1980, he's established the name of Hillspan as one of the big names in steel building frame manufacturing.

DV had followed his elder brothers into the haulage business and after five years driving milk lorries, he came off the road to take charge of the workshops. In 1946, he'd become a founder director - together with his brother Jack - in the newly created Hill & Sons (Botley & Denmead) Ltd. However, when involved in clearing the wartime storage depots, he spotted the potential in selling temporary ex military buildings (as pictured below) and decided to leave the vehicles behind.

Assisted by his wife Marion, DV developed the fledgling concern from an idea to a company with 150 on the workforce and a turnover (in 1980) of £5 million. Hillspan supplied steel for agricultural and industrial buildings both in the UK and overseas. He was a Freeman of the City of London and heavily involved in the charity field. His extensive work on the Royal Estates saw the companies of Hillspan, Hill Construction, Hillspan Engineers, Farmpak Buildings and Hill-Hosier Ltd awarded the Royal Warrant.

Dudley died in July 1999 at the age of 85.

Amongst the projects which Hillspan were involved in was the Queen Elizabeth II Activities Centre for the disabled at Bursledon (bottom) and the heliport at Southampton (top). DV is seen being introduced to Her Majesty The Queen Mother when she visited his stand at the Royal Show, Stonleigh in the early 1970s. He's also seen, on the right of the group shot stood next to Earl Mountbatten of Burma at the Heliport's official opening in January 1969. The figure speaking is the Mayor, Alderman HL Davies while the Sheriff, Alderman Mrs HK Johnson looks on.

HILL & SONS

(BOTLEY & DENMEAD) LTD

Hill

HEAVY HAULAGE SPECIALISTS

Registered Office:

HILLSONS ROAD . BOTTINGS INDUSTRIAL ESTATE . BOTLEY . SOUTHAMPTON SO30 2GE

The idea of compiling a Hills of Botley fleet list goes to Phil Moth, a long-standing fan of the company. He originated the first input although he credits John Mollett with some of the information. Paul Hancox has helped with details from the 1970s whilst other additions have been made from the Hill records and photograph collection. I make no claims that what follows is comprehensive (or even without the odd error - typing or otherwise) as all manner & number of vehicles has passed through the Hill premises at Botley, Denmead, Meonstoke, Petersfield, Reading and London while not forgetting Rotterdam & Paris.

Reg	Make	Type		Reg	Make	Type
PX 7107	Dennis	4 wheel flat		EAM 568	Thornycroft	4 wheeler
RV 2305	Dennis	4 wheel flat		ECK 259	Leyland	8 wheeler
EM 5674	Seddon	4x2 artic unit		EBK 522	Dennis	4 wheel cattle box
JE 8692	Foden	4 w. milk tanker		EDW 169	Dennis	4 wheeler
PO 9639	Morris Commercial	4 wheel tilt body		EHO 431	Thornycroft	4 wheel flat
ACG 636	Ford	4 wheel cattle box		EHO 504	ERF	4x2 artic unit
AFA 545	Seddon	4 wheel flat		EHO 685	ERF CI5	4 wheel flat
AUX 39	Foden DG	4x2 artic unit		EMJ 932	Thornycroft	4 wheeler
BDJ 946	Seddon Mk.5	4x2 artic unit		EMR 491	Bedford OYD	4 wheeler
BBK 553	Albion	4 wheel cattle box		EPM 528	Seddon	4 wheel cattle box
BBK 875	Leyland Lynx	4 wheel flat		ETP 593	Dennis	4 wheel cattle box
BRD 712	AEC M / Major	8 wheel flat		ETR 166	Ford	4 wheeler
CAA 23	Dodge	4 wheel flat		EYW 15	Scammell	4x2 artic unit
CCB 141	Leyland Octopus	8 wheel flat		EYF 281	Scammell	4x2 artic unit
CHO 29	AEC Matador	4x4 tractor		FAA 990	Bedford O	4x2 unit
CMR 16	Austin K3	4 wheeler		FBK 20	Albion	4 wheel flat
COT 794	Bedford O	4x2 artic unit		FBK 326	Morris	4 wheel cattle box
CTP 204	Dodge	4 wheel flat		FDD 650	Albion	4 wheeler
CTP 336	Dennis	4 wheel flat		FOR 16	Dennis	4x2 artic unit
CTP 496	Dennis	4 wheel cattle box		FOR 310	GMC	6x6
CTP 512	Dennis	4 wheel flat		FOR 547	Foden DG	4 wheel tractor
CTP 795	Dennis	4 wheel flat		FOR 829	AEC Matador	4x4 tractor
CTT 246	Foden DG	4 wheel tractor		FOT 846	Bedford	4x2 artic unit
DBK 74	AEC	4 wheel flat		FOU 59	Dennis	Cesspit emptier
DCR 582	Vulcan	4 wheel flat		FOW 401	Seddon	4 wheel flat
DOU 585	Leyland	4 wheeler		FOW 518	Seddon	4 wheel flat
DRN 56	Ford Thames	4 wheeler		FOW 872	Seddon	4x2 artic unit
DRV 648	Dennis Horla	4x2 artic unit		FRU 269	Bedford	4x2 artic unit
DRV 978	Morris	4 wheeler		FWV 171	Albion CX1L	4 wheeler
DTP 112	Dennis	4 wheel cattle box		FXE 37	Scammell	4 wheel tractor
DTR 676	Seddon	4 wheel flat		GAA 311	AEC	6 wheel cattle box
DWV 452	Bedford OLBD	4 wheeler		GAW 743	Leyland Hippo	6 wheel flat
EAD 586	Albion KL127	4 wheeler		GBK 498	Albion	6 wheel flat
				GCJ 940	ERF	4 wheel tanker
				GHO 29	AEC Matador	4x4 tractor
				GHO 246	Bedford	4x2 artic unit
				GLD 666	Scammell	4x2 artic unit
				GLK 782	Bedford	4 wheel tanker
				GOR 903	Dennis Horla	4x2 artic unit
				GOR 919	Dennis Pax	4 wheel cattle box
				GOT 452	Scammell	4x2 artic unit

GVJ 349	Leyland	4 wheel cattle box
GXV 506	Dennis	4 wheeler
GYF 281	Scammell	4x2 artic unit
GYF 492	Scammell	4x2 artic unit
GYF 601	Dennis	4 wheel flat
GYF 627	Dennis	4 wheel flat
GYF 638	Dennis	4 wheel flat
GYF 641	Dennis	6 wheel tanker
GYF 660	Dennis	4 wheel flat
GYF 675	Dennis	4 wheel flat
GYF 681	Dennis	4 wheel flat
GYF 684	Dennis	4 wheel flat
GYF 695	Dennis	6 wheel tanker
GYF 697	Dennis Max	4 wheel flat
HCG 761	Thornycroft	4 wheel cattle box
HCR 207	Seddon	4x2 artic unit
HCR 457	Seddon	4 wheel flat
HDD 887	Leyland	6 wheel tipper
HGT 452	Scammell	4x2 artic unit
HHO 906	Seddon	4 wheel cattle box
HHO 907	Seddon	4 wheel cattle box
HHO 908	Seddon	4 wheel cattle box
HOR 225	Bedford	4x2 artic unit
HPX 407	Quad	4 wheel breakdown
HRJ 174	Ford Thames	4 wheel van
HRU 413	Commer	4 wheeler
HRV 695	Leyland Comet	4x2 artic unit
HTR 38	Seddon	4x2 artic unit
JAA 494	Dennis Max	4 wheel flat
JAA 861	Dennis	4 wheel flat
JBK 295	Leyland Comet	4x2 artic unit
JCR 691	Thornycroft	4 wheeler
JCR 823	Thornycroft	4 wheeler
JDF 813	Seddon	4 wheeler
JFN 27	Ford	4 w. mobile w/shop
JJR 530	BMC	4 wheel tipper
JRD 149	Bedford J	4 wheel horse box
JTR 168	Ford	4 wheel flat
JTR 551	Ford	4 wheel flat
JTR 733	Seddon	4 wheel flat
JXE 746	Scammell	4x2 artic unit
JXM 806	Bedford	4 wheel flat
JXV 483	Dennis	4 wheel flat
KCR 702	Land Rover	4 wheeler
KGH 20	Austin	4 wheel flat
KGH 23	Austin	4 wheeler
KGH 39	Austin	4 wheeler
KGH 41	Austin	4 wheeler
KGH 80	Austin	4 wheeler
KGH 98	Austin	4 wheeler
KGW 637	Morris	Tower wagon
KGY 659	Leyland Comet	4 wheel flat
KGY 662	Leyland Comet	4 w. tackle wagon
KJN 208	Austin	4 wheel tipper
KKJ 469	Seddon	4 wheel flat
KLA 630	Maudslay	4 wheel flat
KLC 63	Austin	4 wheeler
KLC 857	Dennis	4 wheel flat
KOA 978	Leyland Beaver	4 wheel flat
KOA 988	Leyland Beaver	4 wheel flat
KOW 185	Seddon	4 wheel flat
KOW 847	Trojan	4 wheel van
KPE 144	Dennis	4 wheel van
KPF 145	Dennis	4 wheel van

KPO 462	Fordson Major	tractor / loader
KTP 380	Dennis	4 wheel van
KTR 522	Seddon	4x2 artic unit
KWX 965	Seddon	4x2 artic unit
KXT 146	Maudslay	4 wheel flat
KXT 872	Scam. Pioneer	6x4 tractor
KXV 873	Austin	4 wheel tanker
KXY 421	Dodge	4 wheeler
KXY 712	Austin	4 wheel tanker
KXY 730	Albion	4 wheel tanker
KYF 372	Austin	4 wheel tanker
KYK 225	Seddon	4 wheel flat
LBP 277	Seddon	4 wheeler
LLW 955	Leyland Comet	4 wheeler
LKX 230	AEC Matador	4x4 tractor
LOR 712	Leyland Comet	4x2 artic unit
LOM 118	Karrier	4 w. gully emptier
LOM 120	Karrier	4 w. gully emptier
LOW 37	Trojan	4 wheel van
LOW 264	Seddon	4x2 artic unit
LPC 865	Thornycroft	4 wheel tipper
LRB 767	Leyland	6 wheel tanker
LRU 458	Albion Chieftain	4 wheel flat
LRV 315	BMC	4 wheel flat
LRV 316	BMC	4 wheel flat
LTK 969	Seddon	4x2 artic unit
LTK 970	Seddon	4x2 artic unit
LTR 7	Seddon	4x2 artic unit
LTR 8	Seddon	4x2 artic unit
LTR 236	Seddon	4x2 artic unit
LUC 680	Dennis	4 wheel tanker
LWV 539	Albion	4 wheeler
LYL 443	Dennis	4 wheel tanker
LYL 445	Dennis	4 wheel tanker
LYL 453	Dennis	4 wheel tanker
LYL 478	Dennis	4 wheel tanker
LYP 36	Bedford	4 wheeler
LYP 516	Dennis	4 wheel tanker
MED 584	AEC Mercury	4 wheel flat
MJT 123	Seddon	4x2 artic unit
MLE 403	Maudslay	4 wheel flat
MLE 887	Scammell	4x2 artic unit
MLJ 161	Ford Thames	4 wheel flat
MOA 61	Seddon	4x2 artic unit
MOG 568	Dodge	4 wheel flat
MOT 936	Ford	4 wheel meat van
MOU 155	Ford Thames	4 wheel flat
MRV 196	Bedford	4 wheel flat

Reg	Make	Type
MRX 354	Commer	4 w. (2 stroke)
MUH 21	Scammell	4x2 artic unit
MXB 210	Bedford S Type	4x2 artic unit
MXV 869	Seddon	4 wheeler
NAA 401	Bedford	4 wheel cattle box
NGX 819	Seddon	4 wheel flat
NLR 564	Leyland Octopus	8 wheel flat
NNM 311	Dennis	6 wheel flat
NRO 801	Foden DG	4x2 artic unit
NTR 773	Trojan	4 wheel van
NTR 883	AEC Mercury	4x2 artic unit
NTW 410	Leyland Steer	6 wheel flat
OCG 459	Commer	4 wheel flat
OGT 750	Albion	4 wheel tanker
OHO 934	Commer	4 wheel flat
OHR 10	BMC	4 wheeler
OJJ 700	Leyland Octopus	8 wheel cattle box
OLU 281	Scammell	4x2 artic unit
OOR 130	Albion	4x2 artic unit
OOR 318	Commer	4 wheel flat
OOR 584	Bedford	4 wheel flat
OOR 585	Ford Thames	4 wheel flat
OOT 207	Bedford	4 wheel flat
OOT 663	Ford Thames	4 wheel flat
OTC 633	Seddon	4x2 artic unit
OTD 947	Seddon	4x2 artic unit
OVJ 715	Ford Thames	4 wheel van
OXT 491	Seddon	4 wheel flat
PAA 469	Ford	4 wheel flat
PAA 805	Albion	4x2 artic unit
PBK 36	ERF	4 wheeler
PBL 876	Ford Thames Trader	4 wheel tipper
PCG 368	Albion	4x2 artic unit
PCG 481	Bedford	4 wheeler
PCR 412	Seddon	4 wheeler
PEL 89	Albion Chieftain	4 wheeler
PEL 294	Ford	4 wheel van
PHX 395	Austin	4x2 artic unit
PKK 681	Fordson	road roller
PKM 896	Foden	8 wheel flat
PLA 397	Scammell	4x2 artic unit
POT 663	Ford Thames	4 wheel flat
PTG 926	Ford Thames	4 wheel flat
PUC 472	Scam. Constructor	6x6 tractor
PUC 474	Scam. Constructor	6x6 tractor
PXV 308	AEC Mercury	4 wheel flat
PXV 84	Leyland	tanker
RAA 397	Ford Thames 4D	4 wheel flat
RAA 399	Ford Thames 4D	4 wheel flat
RAA 823	Commer	4 wheel flat
RBK 286	Albion	4x2 artic unit
RBK 619	Albion	4 wheeler
RBY 194	Ford Thames 4D	4 wheel van
RHW 780	Trojan	4 wheeler
RHY 768	Trojan	4 wheeler
RLW 962	BMC	4 wheel tipper
RLW 967	Austin	4 wheel flat
RLW 968	BMC	4 wheel tipper
ROT 194	Ford Thames 4D	4 wheel flat
ROT 972	Foden DG	6 wheel tractor
RTK 521	Bedford	4 wheel tipper
RTO 715	Seddon	4x2 artic unit
RUP 900	Scam. Mountaineer	4x4 tractor
RJT 330	Bedford TK	4 wheel tipper
SCG 196	Ford Thames Trader	4 wheel flat
SGK 320	BMC	4 wheel cattle box
SLP 941	Bedford	4 wheel tanker
SLP 943	Bedford	4 wheel tanker
SMF 236	Bedford	4 wheel van
SPR 664	Seddon	4 wheel tanker
SPT 600	Scam. Constructor	6x6 tractor
SRX 529	Ford Thames Trader	4 wheel flat
SRX 530	Ford Thames Trader	4 wheel cattle box
STH 403	Ford Thames	4 wheeler
SUU 523	Dennis	4 wheel tanker
SUU 544	Dennis	4 wheel tanker
SXC 139	Austin	4 wheel tipper
SXC 178	Foden	8 wheel tanker
SXC 320	Foden	4 wheel tanker
SXC 352	Scammell	4x2 artic unit
SXC 355	Scammell	4x2 artic unit
SXT 377	Commer	4 w. flat (2 stroke)
TAX 280	Commer	6 w. flat (2 stroke)
TBM 494	Ford	4 wheeler
TCG 873	Bedford	4 wheeler
TJJ 643	Scammell	4x2 artic unit
TOR 124	Ford	4 wheeler
TOT 297	Scam. Explorer	6x6 tractor
TTR 703	Ford	4 wheeler
TWR 575	AEC	tanker
TYH 811	AEC	tanker
TYO 678	AEC	tanker
UDF 946	AEC	6x6 tractor / crane
UEL 477	Ford Thames	4 wheel cattle box
ULY 889	AEC	tanker
ULT 128	Morris	4 wheel van
UTR 166	Ford Thames Trader	4 wheel flat
UTR 234	Ford Thames Trader	4 wheel flat
UTR 257	Ford Thames Trader	4 wheeler
UTR 586	Ford Thames Trader	4 wheel flat
UXC 957	Leyland Beaver	4x2 artic unit
UYL 576	Bedford	4x2 artic unit
VME 693	Leyland Beaver	4 wheel van
VTR 202	Ford Thames Trader	4 wheel cattle box
VTR 353	Ford Thames Trader	4 wheel cattle box
VTR 240	Ford Thames Trader	4 wheel cattle box
VUW 825	Leyland	4x2 artic unit
WPJ 526	Ford Thames 4D	4 wheel flat
XMT 45	Scammell	4x2 artic unit
XMT 586	Thornycroft Antar	6x4 tractor
XNM 477	Scammell	4x2 artic unit
XPA 519	Ford Thames 4D	4 wheel flat
XRT 521	BMC	4 wheel tipper
XXW 534	Seddon	4x2 artic unit
YCR 507	Ford Thames Trader	4 wheeler
YXC 493	Bedford TK	4x2 artic unit
YUU 134	Ford Thames Trader	4x2 artic unit
YXC 737	Ford Thames Trader	4 wheel tanker
1049 AD	Leyland	4 wheeler
7584 BP	Commer	4 wheel flat
4722 CR	Ford Thames Trader	4 wheeler

5064 CR	Ford Thames Trader	4 wheeler
5183 CR	Ford Thames Trader	4 wheel van
5382 CR	Ford Thames Trader	4 wheel flat
5722 CR	Ford Thames Trader	4 wheel flat
6619 CR	Ford Thames Trader	4 wheel flat
8413 CR	Ford Thames Trader	4 wheel cattle box
1247 PO	Albion Clydesdale	4 wheel flat
1502 PG	Ford Thames Trader	4x2 artic unit
1597 TR	Ford Thames Trader	4 wheel cattle box
1599 TR	Ford Thames Trader	4 wheel cattle box
1601 TR	Ford Thames Trader	4 wheel cattle box
3592 TR	Ford Thames Trader	4 wheeler
2919 VB	Bedford TK	4x2 artic unit
4237 VP	Ford Thames Trader	6 wheeler
4898 VW	Ford Thames Trader	6 wheel flat
12 ABJ	Bedford	4 wheel tipper
24 AHU	Ford Thames	4 wheeler
483 APO	Seddon	4x2 artic unit
662 APO	Dodge	4x2 artic unit
663 APO	Dodge	4x2 artic unit
921 ARM	Ford Thames Trader	4 wheel cattle box
874 AUU	Scam. Junior Constructor	6x4 tractor
824 BMH	Seddon	4x2 artic unit
825 BMH	Seddon	4x2 artic unit
142 BUT	Foden S21	6x4 artic unit
614 BXB	Ford Thames Trader	4x2 artic unit
6 CCR	Ford Thames Trader	4 wheeler
563 CHO	Ford Thames Trader	4 wheel cattle box
946 CWP	Austin	4 wheel tipper
291 DCG	BMC	4 wheeler
120 DYC	Federal	4x4
230 EPH	Dennis	4 wheel flat
358 ETN	Scam. Constructor	6x6 tractor
298 FUW	Scam. Mountaineer	4x4 tractor
332 GL?	Guy Warrior	6 wheel flat
703 GOU	Ford Thames Trader	4x2 artic unit
741 GPG	Ford Thames Trader	4 wheel flat
284 HHU	Leyland	4 wheel flat
187 JKM	BMC	4 wheel tanker
189 JKM	BMC	4 wheel tanker
190 JKM	BMC	4 wheel tanker
191 JKM	BMC	4 wheel tanker
287 JME	Foden	6 wheel tractor
707 JOR	Ford Thames Trader	4x2 artic unit
122 MMX	Ford Thames 4D	4 wheel van
840 NHO	Albion Clydesdale	4x2 artic unit
767 SRR	Ford Thames	4 wheel cattle box
774 YMD	Dodge	4 wheel flat
CMH 978A	Ford	4 wheeler
AAB 797B	Ford Thames Trader	4x2 artic unit
AOR 325C	Albion Clydesdale	4x2 artic unit
BAA 387C	Albion Clydesdale	4x2 artic unit
FLT 758C	Guy Invincible	4x2 artic unit
DCG 887C	Ford D Series	4 wheel tanker

HOM 158D	Bedford TK	4 wheel tipper
HOR 924E	Scammell Pioneer	6x4 recovery tractor
KRW 591E	Guy Big J	4 wheel artic unit
KRE 15F	Guy Big J	4 wheel artic unit
LBO 632F	Guy Big J	6x4 artic unit
MKG 631F	Guy Big J	6x4 artic unit
MNV 510F	Scam. Handyman	4x2 artic unit
NOT 30F	Hill (Guy Invincible)	6x2 artic unit
NOT 79F	Scammell	6 wheel tractor
PYH 546F	Foden S36	6x4 artic unit
SJD 801F	Scam. Contractor	6x4 tractor
SYO 376F	Scam. Contractor	6x4 tractor
TBJ 351F	Scam. Trunker III	6x2 artic unit
RAA 418G	Guy Big J	4x2 artic unit
ROT 428G	Hill (Guy Invincible)	6x2 artic unit
DBF 718H	ERF LV	6x4 artic unit
JPT 553H	Volvo F88	6x4 artic unit
TOU 997H	Hill (Guy Big J)	6x2 artic unit
TRL 924H	Scam. Contractor	6x4 tractor
UAA 896H	Guy Big J	4x2 artic unit
UOT 890H	Guy Big J	4x2 artic unit
VAA 380H	Guy Big J	4x2 artic unit
VOU 530H	Guy Big J	4x2 artic unit
SAR 457J	Scammell Crusader	6x4 artic unit
SNB 431J	Scammell Crusader	6x4 artic unit
TUX 450J	Foden S40	6x4 artic unit
UFU 937J	Foden S40	6x4 artic unit
FAA 847K	Hill (Guy Invincible)	6x2 artic unit
EWW 259L	Scammell Crusader	4x2 artic unit
JOG 20L	Scammell Crusader	4x2 artic unit
MOT 397L	Bedford TK	4x2 box van
NVO 203L	Scammell Crusader	4x2 artic unit
XTP 324L	Scammell Crusader	4x2 artic unit
DET 969M	Atkinson Borderer	4x2 artic unit
SUS 684M	MAN	4x2 artic unit
LUH 187P	MAN 19.330	4x2 artic unit
LVS 83P	Daf	6 w. artic unit
MRB 229P	Scammell Crusader	6x4 artic unit
NHT 849R	MAN 19.330	4x2 artic unit
RJT 512R	Daf	6 w. artic unit
SUS 684R	MAN	4x2 artic unit
VPO 192T	MAN 19.330	4x2 artic unit
APX 595V	Mercedes Benz 307	4 wheel van
YPO 616V	MAN	4x2 artic unit
HBM 529Y	Scammell S26	6x4 artic unit
LTR 689Y	Scammell S24	6x4 artic unit
A390 OPX	Scammell S24	6x4 artic unit
A635 VNV	MAN	6x2 artic unit
A636 VNV	MAN	6x2 artic unit
B810 DWE	Daf 3300	6 w. artic unit
G992 ROD	Scania 143-450	8x4 artic unit
K320 NMW	MAN	6x6 artic unit
T198 PBP	Iveco	8x4 artic unit